NANOTECHNOLOGY SCIENCE AND TECHNOLOGY

MAGNETIC NANOPARTICLES

SYNTHESIS, PHYSICOCHEMICAL PROPERTIES AND ROLE IN BIOMEDICINE

NANOTECHNOLOGY SCIENCE AND TECHNOLOGY

Additional books in this series can be found on Nova's website under the Series tab.

Additional e-books in this series can be found on Nova's website under the e-book tab.

NANOTECHNOLOGY SCIENCE AND TECHNOLOGY

MAGNETIC NANOPARTICLES

SYNTHESIS, PHYSICOCHEMICAL PROPERTIES AND ROLE IN BIOMEDICINE

NORA P. SABBAS

EDITOR

nova publishers

New York

For permission to use material from this book please contact us:
Telephone 631-231-7269; Fax 631-231-8175
Web Site: http://www.novapublishers.com

Library of Congress Cataloging-in-Publication Data

ISBN: 978-1-63117-434-6

Published by Nova Science Publishers, Inc. † New York

CONTENTS

PREFACE

Magnetic nanoparticles (MNPs) are one of the materials of great interest for presenting a unique combination of relevant properties such as high surface area, magnetic behavior and low toxicity, which can find potential use in different processes and applications in areas such as catalysis, data storage, water treatment, drug delivery system, DNA separation, tissue engineering, sensors, hyperthermia, ferrofluids, and as contrast agents in nuclear magnetic resonance (NMR) imaging. This book provides further information on how magnetic nanoparticles are synthesized, their physicochemical properties and the roles MNPs play in biomedicine.

Chapter 1 - Magnetic nanoparticles (MNPs) are one of the materials of great interest for presenting a unique combination of relevant properties such as high surface area, magnetic behavior and low toxicity, which can find potential use in different processes and applications in areas as catalysis, data storage, water treatment, drug delivery system, DNA separation, tissue engineering, sensors, hyperthermia, ferrofluids, and as contrast agents in nuclear magnetic resonance (NMR) imaging. Modification of the particle surface induces changes in some properties such as high magnetization values and stable water dispersion and recent advances in nanotechnology have improved the range of use of MNPs in diagnosis and therapy.

Among the biological uses of MNPs, the author can highlight the biomedical imaging and therapeutic applications. Chemical modifications of the MNPs surface might result in non-toxic and biocompatible nanoparticles for biomedical applications like the development of new targeted nanoparticles for drug delivery systems for specific tissues. As previously mentioned, MNPs are used as a targeted carrier to be available for drug delivery to the specific site under the influence of a guiding magnet for treatment of diseases.

Furthermore, MNPs offer excellent prospects for chemical and biological sensing. In the case of bioanalytical applications, MNPs can be adopted as nanoparticle-modified transducers for use as sensors and biomolecule-nanoparticle conjugates as labels for biosensing and bioassays. In the present review the author will discuss current uses of MNPs in biomedicine and nanomedicine.

Chapter 2 - Carbon–coated magnetite nanoparticles (NPs) were synthetized by the mechanochemical method with hematite as precursor and amorphous carbon as inorganic reductor. After 18 hours of milling in an inert atmosphere, a nanocomposite material of magnetite and carbon was obtained.

Structural and magnetic properties of the NPs were investigated by X-ray diffraction (XRD), scanning electron microscopy (SEM), transmission electron microscopy (TEM), energy dispersive X-ray spectroscopy (EDS) and vibrating sample magnetometry. XRD patterns, refined with the Rietveld method, show that magnetite is present in samples milled from 6 hours onward and that after milling for 18 hours and annealing in Ar, the sample contains a single crystalline phase. Magnetization curves for samples with different milling times show saturation magnetization values that range from 34.1 emu/g after 1 h to 78.0 emu/g after 18 h. Coercive fields are about 500 Oe for all samples. TEM studies reveal that the samples are made of amorphous carbon clusters with magnetite NPs of 20 nm.

The obtained NPs, associated to electrochemical transducers, show an improved enhancement of the charge transfer for redox processes involving different bioanalytes. Thus, these NPs offer unique properties as a catalyst in biosensing strategies for the electrochemical detection of high-impact markers and the development of theranostics smart-devices for biomedical applications.

Chapter 3 - One of the most rapidly growing and exciting applications of nanotechnology in biomedical research is the development of biocompatible magnetic nanoparticles (NPs) for use in powerful and highly complementary bio-imaging modalities. Biocompatible magnetic NPs should be effective as contrast agents (CAs) in magnetic resonance imaging (MRI), and the unique physical properties of magnetic NPs as well as their conjugation with dyes and/or radioactive elements could lead to the development of the multimodal bio-imaging probes for use in MRI/optical imaging, PET/MRI, and MRI/optical/therapy. Multimodal bio-imaging is quite attractive for the early and precise diagnosis of disease, leading to theranostics, since it combines the advantages of, and provides complementary information from, several imaging modalities.

The metabolism of magnetic NPs strongly depends on their size, but also on individual surface characteristics, like charge, morphology and surface chemistry. Accordingly, the creation of novel magnetic NPs for multimodal bio-imaging can be realized only with the fusion of multiple disciplines, including chemistry, material science, biology, and engineering.

In this chapter, recent advances in the use of magnetic NPs, represented by iron oxide NPs, manganese oxide NPs, and gadolinium oxide NPs, for multimodal bio-imaging are highlighted and summarized.

Chapter 4 - Superparamagnetic iron oxide nanoparticle (SPION) has attracted a great deal of attention in the fields of medicine and biology due to its non-toxicity, large specific surface area, amenability to various surface fuctionalization, and unique magnetic properties. For biomedical applications, the author synthesized two kinds of SPIONs by co-precipitation and polyol methods, designated as *C*- and *P*-SPIONs, respectively. To increase their solubility and stability in an aqueous solution, both of the SPIONs were functionalized with hyperbranched polyglycerol (PG) through ring-opening polymerization of glycidol. Scanning transmission electron microscopy (STEM) and dynamic light scattering (DLS) measurements revealed that the *P*-SPION provided individual nanoparticles without aggregation after the PG fuctionalization. This *P*-SPION-PG was highly soluble not only in pure water (> 40 mg/mL), but also in a phosphate buffer solution (> 25 mg/mL). Such high solubility enabled separation of *P*-SPION-PG according to their size by size exclusion chromatography (SEC). The size-separated *P*-SPION-PG shows a gradual increase in transverse relaxivity (r_2) with increasing particle size. The hydroxyl group in PG also serves as a scaffold for further surface functionalization. For targeted cell labeling, the author functionalized *P*-SPION-PG through multi-step organic transformations (-OH → -OTs (tosylate) → $-N_3$ → -RGD) including click chemistry as a key step to impart targeting specificity by immobilization of cyclic RGD peptide on the surface. The targeting effect was demonstrated by the cell experiments; *P*-SPION-PG-RGD was taken up by the cells overexpressing $\alpha_v\beta_3$-integrin such as U87MG and A549. In addition, near-infrared (NIR) fluorescence dye (IRdye 800CW) was immobilized on the surface of *P*-SPION-PG through amide linkage. The resulting *P*-SPION-PG-IRdye showed good solubility in phosphate buffered saline (PBS) and strong fluorescence in NIR region, which may find application in optical/MR bimodal imaging.

Chapter 5 - Magnetic nanoparticles and nanocomposites have been special focus of high research interest due to their high magnetic moment and bio-affinity surface properties and promising applications in nanoscience, nanotechnology and bio-applications. These properties can be obtained by synthesizing various magnetic nanoparticles and core-shell type magnetic nanocomposites. In this research, the author synthesized high magnetization superparamagnetic nanoparticles and core-shell nanocomposites by chemical and sonochemical method for using as carriers/labels in bio-sensing purpose.

Iron oxide magnetic nanoparticles (NPs) have been synthesized by sonochemical method using inexpensive and non-toxic metal salts as reactants. Transmission electron microscopy (TEM) data demonstrated that the particles were narrow range in size distribution with 11 nm average particle size and spherical in shape. The magnetization curve from vibrating sample magnetometer (VSM) measurement shows that as-synthesized NPs were nearly superparamagnetic in magnetic properties with very low coercivity, and magnetization values were 80 emu/g.

Monodisperse magnetite nanocubes with uniform particle size of about 80 nm have been synthesized in aqueous medium by sonochemical method. The magnetic characterization of the NPs reveals saturation magnetization of 91 emu/g at 5 K for as-synthesized sample and 94.8 emu/g for the sample which annealed at the temperature of 600 °C in a vacuum chamber. However, the saturation magnetization has been observed to decrease with further increase in annealing temperature and this has been attributed to the presence of a thin magnetic dead layer at the surface caused by shape anisotropy distortion and broken exchange bonds, and spin canting on the surface of the particles in addition to formation of a small amount of maghemite phase.

Highly crystalline and monodisperse cobalt ferrite ($CoFe_2O_4$) nanoparticles have been synthesized via rapid one-pot sonochemical techniques and without subsequent calcination. The size of $CoFe_2O_4$ nanoparticle was controlled in the range from 20 to 110 nm based on the solvent medium used in the synthesis process. Furthermore, the evolution from spherical to cubic morphology of cobalt ferrite is achieved by simply changing the solvent medium from aqueous to alcoholic medium. High saturation magnetization (Ms) and high coercivity (Hc) values of 87 emu/g and 1610 Oe, respectively were obtained for the $CoFe_2O_4$ NPs.

For synthesis of core-shell type of silica coated iron oxide magnetic (Fe_3O_4@SiO_2 core-shell) NPs, sono-chemical approach was applied using inexpensive and non-toxic chemicals. TEM data demonstrated that the thickness of silica coating on iron oxide magnetic NPs 10-15 nm in average. The magnetization curve from VSM measurement shows that the magnetization has also been decreased of as synthesized silica coated iron oxide NPs compared to freshly prepared bare iron oxide magnetic NPs, which is also a evidence of synthesizing of $Fe_3O_4@SiO_2$ core-shell NPs.

Besides the synthesis of metal oxide core-shell NPs, high magnetization FeCo nanoparticles with different Fe/Co ratios have been successfully synthesized by surfactant free simple modified polyol method. In this process, polyethylene glycol was used as a solvent media and it has been found to play a key role to act as a reducing agent as well as a stabilizer simultaneously. TEM data suggest that the annealed FeCo nanoparticles are of 50–90 nm in size. The physical Property Measurement System (PPMS) reveals that the $Fe_{60}Co_{40}$ composition among all the samples exhibit highest saturation magnetization of 230.14 emu/g at 5K.

In another study, high magnetic monodisperse NiFe NPs with different compositions have been successfully synthesized polyol method. TEM images displayed formation of a thin oxide layer around the nanoparticles, and confirmed by detection of some oxygen element using EDS measurement. The magnetic properties of the synthesized NiFe NPs samples were measured VSM at room temperature, and the saturation magnetization value was found to be iron content dependent.

Chapter 6 - Materials capable of responding to external stimuli including pH, temperature, magnetic and electric field, undergoing conformational changes are considered to be one of the most exciting and emerging classes of advanced materials receiving considerable scientific interest especially in the biomedical field. Stimuli-responsive polymers in the form of micro- or nanofibers have received great attention during the last 10 years and have been exploited in a diverse range of biomedical applications such as drug delivery, tissue engineering, bioseparation and biosensing. One of the most popular and versatile fiber fabrication methods used for the production of fibers is electrospinning. Its simplicity, cost-effectiveness and applicability not only to pristine synthetic and natural polymers but also to composites, enables the development of polymer-based fibrous nanocomposites *via* the combination of polymers with inorganic nanofillers. Among such nanoadditives, magnetic nanoparticles capable of interacting with an externally applied magnetic field, are particularly attractive owing to their potential biomedical applications including magnetically-triggered drug delivery, magnetic cell seeding, magnetic bioseparation, hyperthermia cancer treatment and contrast enhancement in magnetic resonance imaging. In this chapter, an introductory section on electrospinning and on different parameters influencing this process is initially provided. The different fabrication routes for generating electrospun magnetoactive polymer-based (nano)fibrous materials are briefly discussed and finally the applicability of these materials in the biomedical field including tissue engineering, drug delivery, hypethermia treatment and biosensing is reviewed.

Chapter 7 - Interest to the nanomedicine has increased dramatically during past years. Since nanotechnology has the potential to have a revolutionary impact on cancer diagnosis and therapy, is there a major challenge to produce magnetic nanoparticles (MNPs) suitable for biomedical applications. In this review the main theoretical views on the properties, synthesis, characterization, functionalization, and bioapplication of MNPs in the current context of science are considered. This review examines some of the recent developments in MNPs technology and provides a brief background of their applications and results of *in-vitro* and *in-vivo*, animal and clinical experiments of targeted drug delivery. Finally, some of the recent biological, medical and scientific applications of MNPs are briefly reviewed, and some future trends and perspectives in these research areas will be outlined.

In: Magnetic Nanoparticles
Editor: Nora P. Sabbas

ISBN: 978-1-63117-434-6

Chapter 1

BIOMEDICAL APPLICATIONS OF MAGNETIC NANOPARTICLES: AN OVERVIEW

***G. A. E. Chávez*[1]*, Ll. J. C. Medina*[1]*, C. P. de Melo*[2]*, C. A. S. Andrade*[3] *and M. D. L. Oliveira*[3]**

[1]Programa de Pós-graduação em Ciência de Materiais,
Universidade Federal de Pernambuco,
Recife, PE, Brasil

[2]Departamento de Física,
Universidade Federal de Pernambuco,
Recife, PE, Brasil

[3]Departamento de Bioquímica,
Universidade Federal de Pernambuco,
Recife, PE, Brasil

1. ABSTRACT

Magnetic nanoparticles (MNPs) are one of the materials of great interest for presenting a unique combination of relevant properties such as high surface area, magnetic behavior and low toxicity, which can find potential use in different processes and applications in areas as catalysis (Lu et al., 2007), data storage (Frey et al., 2009), water treatment (Meng et al., 2011), drug delivery system (Anirudhan et al., 2013b), DNA separation (Chiang et al., 2005), tissue engineering (Ito and Honda, 2007), sensors (Baby and Ramaprabhu, 2010), hyperthermia (Alphandéry et al., 2012), ferrofluids (Hee Kim et al., 2005), and as contrast agents in nuclear magnetic resonance (NMR) imaging (Choi et al., 2004). Modification of the particle surface induces changes in some properties such as high magnetization values and stable water dispersion and recent advances in nanotechnology have improved the range of use of MNPs in diagnosis and therapy.

Among the biological uses of MNPs, we can highlight the biomedical imaging and therapeutic applications. Chemical modifications of the MNPs surface might result in non-toxic and biocompatible nanoparticles for biomedical applications like the development of new targeted nanoparticles for drug delivery systems for specific tissues. As previously mentioned, MNPs are used as a targeted carrier to be available for drug delivery to the specific site under the influence of a guiding magnet for treatment of diseases. Furthermore, MNPs offer excellent prospects for chemical and biological sensing. In the case of bioanalytical applications, MNPs can be adopted as nanoparticle-modified transducers for use as sensors and biomolecule-nanoparticle conjugates as labels for biosensing and bioassays. In the present review we will discuss current uses of MNPs in biomedicine and nanomedicine.

Keywords: Magnetic nanoparticles, biomolecules, biomedicine, therapy, diagnosis

2. Magnetic Nanoparticles

In recent years, the nanotechnology has been of great interest in Physics, Chemistry, Science and Engineering of Materials, and Biotechnology, because it is bringing technological breakthroughs (Charles P. Poole and Owens, 2003). Nanotechnology involves the control and manipulation of materials with at least one dimension smaller than 100 nanometers; when the matter is in the nanometric scale has different properties in relation to bulk materials (Charles P. Poole and Owens, 2003).

MNPs are one of the materials of main interest in this area due to their properties as high surface area to volume ratio, magnetic behavior and low toxicity. Thus, diverse applications are discussed in the subsequent topics to introduce the reader to the use of MNPs in biomedicine.

2.1. Types of MNPs Used in Biomedicine

Although the MNPs commonly consist of chemical compounds of magnetic elements such as iron, nickel and cobalt, most utilized are magnetite (Fe_3O_4) and maghemite (γ-Fe_2O_3). In these compounds, when the particle size is below of a certain critical value (20 nm) the nanoparticle (NP) exhibits a superparamagnetic behavior (Tan, 2009) that prevents the agglomeration of particles since they now only become magnetic when an external magnetic field is present.

2.2. Synthesis of MNPs

Several procedures have been extensively used for the preparation of MNPs and some examples are discussed here.

- *Chemical co-precipitation* (Alibeigi and Vaezi, 2008; Lee et al., 2004b; Liu et al., 2002): this method consists in adding a base (typically NaOH or NH_4OH) to a solution containing a mixture of two sources of iron (chlorides, sulfates and nitrates are the most common iron (III) and iron (II) sources used). The main advantage of this method of producing MNPs is that the easy synthetic procedures involved can be carried scaled up to industrial scale. In the literature, this type of synthesis can be found in some variants, such as the addition of citrate ions to the reaction system to obtain a better control of the particle size (Liu et al., 2004b) since through complexation these ions affect growth of iron oxide nanoparticles.
- *Thermal decomposition in organic media* (Roca et al., 2007; Tartaj et al., 2005): this method consists in the decomposition of iron organic precursors at elevated temperatures, after their injection into a hot solution of organic solvents in the presence of surfactants. Typically, cuprates or carboxylates may be used as iron organic precursors, while oleic acid is the surfactant, and phenyl ether as the organic solvent. This technique allows the control of particle size, and the synthesized particles have high crystallinity.
- *Microemulsions* (Tartaj et al., 2005; Zhou, 2001): these systems are obtained by dispersing two immiscible liquids, such as oil and water, in the presence of a surfactant. The surfactant provides confinement effects that limits the particle nucleation, growth and agglomeration.

- *Condensation by vaporization* (Tartaj et al., 2005): this is a process whereby a solid is obtained by spraying of a solution in a series of reactors. Where after the evaporation of the solvent in the solute-solvent droplets the solute experiences condensation, which is followed by drying and heating treatment.
- *Oxidation-precipitation* (Deng et al., 2003): this is a method similar to chemical co-precipitation, the main difference in oxidation-precipitation is that just one source of iron is used in the synthesis of MNPs, such as $FeSO_4$.

2.3. Properties of MNPs

MNPs, mostly magnetite (Fe_3O_4) and maghemite (γ-Fe_2O_3), have been receiving great attention due to the possibility of uniting their unusual magnetic properties in nanoscale technological applications in engineering, catalysis, mineralogy, biology and medicine. When the grain size of the particles is less than 20 nm, the particles exhibit a superparamagnetic behavior (Tan, 2009). The superparamagnetic particles are magnetized when a magnetic field is applied and lose their magnetic properties once the field is turned off. Another interesting characteristic of MNPs is that become heated when subjected to an external magnetic AC field. The origin of this heating is consequence of the energy loss during the demagnetization process.

The demagnetization is due to a reduction in the anisotropy energy, this energy holds the magnetization along of a specific orientation in the particle, and can be expressed by the product of the constant of anisotropy, k, and volume, V, and this product can be described as the energy barrier required to reverse the magnetization of the particles. When the volume of a particle is small, kV becomes so small that the thermal energy can overcome the forces of anisotropy and reverse the magnetization of the particle (Spaldin, 2003). The superparamagnetism is a behavior similar to paramagnetism, the difference is that superparamagnetic materials have a higher magnetic susceptibility.

3. Surface Modification of MNPs for Biotechnological Applications

One of the major problems of using MNPs in any application is that tend to agglomerate, to overcome this problem has been performed much effort in order to find compounds which can modify its surface and prevent the agglomeration. Some of the compounds that have been used are carboxylates, phosphates, sulfates, silica, gold, dextran, polysaccharides, polyvinyl butyral, poly(methyl methacrylate), polyesters, chitosan, polyethylene glycol, and polyvinyl alcohol.

When the surface of MNPs is modified, the magnetic properties are maintained and can be introduced new features such as stability, biocompatibility and biodegradability. Furthermore, the modifying compounds can have reactive groups such as amino, aldehyde, hydroxyl and carboxyl groups, which can be bound to biomolecules as DNA, protein, enzyme and antibody increasing the possible biomedical applications of the MNPs.

Georgelin et al. (Georgelin et al., 2008) reported the use of organophosphorous ligand terminated by an amine group to functionalize γ-Fe_2O_3 NPs and subsequently immobilize an enzyme onto particles surface in a covalent mechanism. In that study, the enzymatic activity of the bounded enzyme was evaluated in terms of the classical Michaelis-Menten kinetics. The authors observed that the affinity of the bounded enzyme for the substrate is preserved and is not affected by the high amount of enzymes on the particle surface. Moreover, the enzymatic activity slightly decreased when compared to that of the free enzyme.

Currently, research strategies in biotechnological application are related to the synthesis of biocompatible MNPs. The approach for surface functionalization of MNPs with boronic acid and their use for the covalent immobilization of adenosine is an advance in nucleic base identification, separation and genosensor strategy (Pham et al., 2010c). Pham et al. (Pham et al., 2010b) reported a facile approach for surface functionalization of MNPs with boronic acid for covalent immobilization of adenosine. The surface functionalization of MNPs modified with a thin gold layer was carried out using 4-mercaptophenylboronic acid through the Au-S chemistry. The immobilization was obtained via a strong covalent interactions of the boroester linkages, which are formed between free hydroxyl groups of adenosine and boronic acid (Pham et al., 2010b). After the modification process, a good separation of adenosine-immobilized on MNPs was obtained under an external magnetic field.

Also, some studies have discussed the use of silane agent as an element to modify the iron oxide nanoparticles surface, improving the biocompatibility as well as high density of surface functional end groups to connect to other metal, polymer or biomolecule (Ma et al., 2003; Shen et al., 2004). In general, while silane-coated iron oxide NPs still maintains the physical characteristics of naked NPs with a high saturation magnetization, a decrease in the degree of functionalization of MNPs with amino group, silica, polymer, various surfactants or other organic compounds is usually desired in order to achieve better physical and chemical properties. A simple strategy to obtain functionalized silica shell-magnetic nanoparticles has been developed as a useful tool to immobilize enzymes (Sulek et al., 2010). Sulek et al. (Sulek et al., 2010) performed the surface modification of maghemite (γ-Fe_2O_3) NPs with silica followed by glutaraldehyde activation in order to improve the protein immobilization.

4. Biomedicine Applications of MNPs

4.1. Bioseparation

The separation of a target compound is an important process in several areas, for example, environmental science (Bhaumik et al., 2011), food (Cao et al., 2012), biomedicine (Li et al., 2011), and among others. In biomedicine, it has a special interest exists for the separation of biomolecules such as proteins, enzymes, nucleic acids and cells. There is a wide variety of techniques commonly used to separate these biomolecules such as chromatography, precipitation, filtration, centrifugation, dialysis, etc. However, these separation methods have some limitations as being time-consuming, or requiring expensive instrumentation and specialized operators (Cao et al., 2012). For this reason, in the last years much effort has been dedicated to the development of new and more efficient techniques of separation. Among the several advantages that magnetic separation has when compared to conventional separation methods once can point out that they are quicker, simpler and cheaper, while not requiring use of sophisticated equipment (Laurent et al., 2008). Magnetic separation consists in incorporating MNPs into a medium with the objective of recovering a target biomolecule by applying an external magnetic field to the system (Figure 1). For this, usually the MNPs must be functionalized with compounds that have affinity to the target biomolecule.

In bioseparation, the MNPs' surface is modified with different types of organic and inorganic compounds to prevent the self-aggregation of the MNPs. Furthermore, chemical modification allows the incorporation of functional groups on the particle surface that can interact selectively with the target molecules. This interaction can occur through a covalent bonds between the functionalized surface of the MNP and the target molecule, or by a direct electrostatic interaction between them.

Below, we discuss some of the investigations that have been dedicated to the use of MNPs in magnetic bioseparation, while in Table 1 we list some of the magnetic compounds used for the separation of proteins, enzymes, nucleic acids and cells.

Li et al. (Li et al., 2011) synthesized magnetic mesoporous silica nanoparticles (MMSN) with a core-shell structure, with a Fe_3O_4 particle in the nucleous and a mesoporous silica shell. They studied the DNA adsorption process as a function of time in a chaotropic salt solution and found that the MMSN adsorbed a high amount of DNA (121.6 mg/g), mainly on its external surface and into the mesopores. The desorption studies were performed through changes in the system temperature, and from these it was possible determine the amount of DNA absorbed on the external surface and into the mesopores (about of 89.5% was captured into the mesopores). It is worth to note that DNA may generate more hydrogen bonds into the mesopores than on the external surface. Due to the high surface area and porous structure, the studied system can be used in other applications, such as controlled drug release. Ashtari et al. (Ashtari et al., 2005) also used Fe_3O_4 NPs coated with silica, but in that work the DNA was absorbed only on the particle surface, which was modified with amino groups and the DNA target was immobilized onto particles through of streptavidin conjugation using covalent bonds.

Liu et al. (Liu et al., 2004a) reported the use of Fe_3O_4 NPs coated with silica functionalized with an amino-silane coupling agent for protein immobilization. Fe_3O_4/silica had a diameter of 50-80 nm in diameter and exhibited a superparamagnetic behavior. Bovine serum albumim (BSA) was covalently immobilized on the NPs surface using glutaraldehyde method. The authors also investigated the influence of pH (which was varied from 3 to 9) on the BSA immobilization capacity, and found that the immobilization was maximum (86 mg/g) at pH 5.

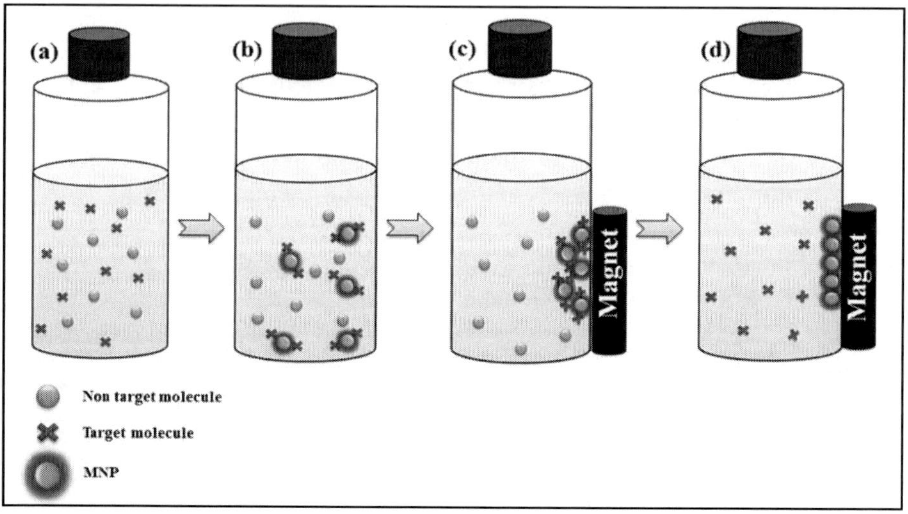

Figure 1. Illustrative representation of magnetic bioseparation of a target biomolecule. System containing target and non-target molecules (a), incorporation of the MNPs to system and their interaction with the target biomolecule (b), application of an external magnetic field to separate the MNPs containing the target biomolecule (c), and re-dispersion of the MNPs/Biomolecules to release the biomolecules of MNPs (d).

Scarberry et al. (Scarberry et al., 2010) synthesized superparamagnetic $CoFe_2O_4$ NPs functionalized with ephrin-A1 mimetic peptide that were then used to remove ovarian cancer cells from peritoneal fluid. Ephrin-A1 has a high affinity to the EphA2 receptor that is present in the ovarian cancer cell. They used flow cytometry to determine the number of cancer cells removed and demonstrated that the metastasis may be abated due to the removal of cancer cells, with a subsequent increase in the survival rates of the ovarian cancer patients.

Some authors reported the preparation of core-shell MNPs of Fe/Au functionalized with boronic acid, with Fe core and Au shell for covalent immobilization of adenosine (Pham et al., 2010a). The authors observed that MNPs had saturation magnetization (Ms) values of 18 emu/g and 11 emu/g before and after of the adenosine immobilization, respectively (Pham et al., 2010a). This decrease in the Ms indicates the presence of adenosine in the surface of the functionalized MNPs. Their magnetic characteristic allowed the separation of MNPs/adenosine of the medium through the application of an external magnetic field. The authors (Pham et al., 2010a) established that this system can be used for the immobilization and separation of other biomolecules such as glucose, catechol and fructose.

Li et al. (Li et al., 2010a) used Fe_3O_4 NPs functionalized with chitosan for covalent immobilization and separation of yeast alcohol dehydrogenase from *Saccharomyces cerevisiae.* They performed various studies on the immobilization of the enzyme, such as immobilization time, enzyme concentration and the influence of pH, and found that these parameters were 120 min, 0.23 mg/mL and pH 7.4, respectively, the NPs reached their maximum efficiency in enzyme immobilization. They also studied the effect of pH and temperature on the activity of the free and immobilized enzyme and was obtained an activity recovery of 65% for immobilized enzyme at 30°C and pH 7.4.

Table 1. Magnetic compounds reported in the separation of proteins, enzymes, nucleic acids and cells

Magnetic compound	Modification of the MNP	Target Biomolecule	Type of biomolecule	Ref.
Fe_3O_4	Mesoporous silica	DNA	Nucleic acid	(Li et al., 2011)
Fe_3O_4	Silica functionalized with amino groups	DNA	Nucleic acid	(Ashtari et al., 2005)
Fe_3O_4	Silica functionalized with an amino-silane coupling agent	BSA	Protein	(Liu et al., 2004a)
$CoFe_2O_4$	Ephrin-A1m	Ovarian cancer cells	Cell	(Scarberry et al., 2010)
Fe	Au functionalized with boronic acid	Adenosine	Nucleoside	(Pham et al., 2010a)
Fe_3O_4	Chitosan	Yeast alcohol dehydrogenase	Enzyme	(Li et al., 2010a)
Fe_3O_4	Polyarginine	BSA	Protein	(Lai et al., 2012)
Fe_3O_4	PVB PMMA	BSA Lysozyme	Protein Enzyme	(Käppler et al., 2008)
$Ni_{0.5}Zn_{0.5}Fe_2O_4$	-	BSA	Protein	(Jiang et al., 2011)
Fe_3O_4	Chitosan	BSA	Protein	(Ghaemy and Naseri, 2012)
Fe_3O_4	Silica decorated with Zn	BSA	Protein	(Bele et al., 2008)
Fe_3O_4	Silica	Glucose oxidase	Enzyme	(Ashtari et al., 2012)

4.2. Hyperthermia

Cancer (CA) is a disease that involves the uncontrolled growth of abnormal cells, which can invade and destroy nearby tissues. These abnormal cells can spread to other parts of the body and affect them. CA affect people of all ages, but the risk is greater with increasing age of the person. According to the World Health Organization (Organization, January 2013), CA is a leading cause of death worldwide; in 2008 it was responsible for 7.6 million deaths (around 13% of all deaths). The main types of CA that cause more deaths for year are lung, liver, colorectal, stomach, breast and cervical CA.

Currently, there are several cancer treatments such as surgery, radiotherapy, chemotherapy and hyperthermia which aim is to cure the disease or considerably prolong life of the patients (Organization, January 2013). In special, hyperthermia has attracted significant attention since it can be used in conjunction with another type of CA treatments to make them more effective (Wust et al., 2002). This treatment consists in to expose the affected tissue to slightly higher temperatures (40 °C-45 °C) that of the normal body metabolism (37 °C) so as to damage or kill the cancer cells (Srirajaskanthan and Preedy, 2012). Since tumor cells are more temperature sensitive than healthy cells, hyperthermia may cause their apoptosis or make them more susceptible to be killed by other complementary treatments, as radiation or chemotherapy. Different methods of hyperthermia are currently under investigation, such as local, regional, and whole-body hyperthermia (Institute, 2011).

In this section, we have particular interest in local magnetic hyperthermia, which is a therapeutic procedure used to increase the temperature of a specific region in the body affected by malignant agents. The change in the temperature is induced by the use of MNPs that are injected into affected region and by the action of an external magnetic field it is possible to selectively kill the malignant agents (Figure 2) (Tartaj et al., 2003). The heat generation of MNPs is due to loss mechanisms such as Brownian (heat due to friction emerging from particle oscillations) and Néel (heat due to rotation of the magnetic moment) effects (Cherukuri et al., 2010).

Magnetic hyperthermia has been used in both animal and humans to treat a series of different CAs located at different regions of the body such as prostate (Johannsen, 2010), breast (DeNardo et al., 2007), head and neck (Qun Zhao, 2012). Some factors must be considered in order to have an efficient use of MNPs in hyperthermia, such as the heating capacity of the MNP, its specific absorption rate (SAR) and the type of modification implemented on the surface of the MNP. The first factor allows the temperature control generated by the MNPs inside of body when the magnetic field is applied: the temperature has to be high enough and be kept for a certain time to damage or kill CA cells, but must not exceed the 45 °C so as not to cause damage to the healthy cells (Ban et al., 2011). To attain a better temperature control, several researchers have synthesized MNPs with Curie temperatures (T_c) in the 40°C - 45°C range. The T_c is the temperature above which a ferromagnetic material switches to a paramagnetic behavior: since no more heat is generated when the T_c is reached, a self-regulation of the temperature is attained with the use of these optimized MNPs (Ban et al., 2011).

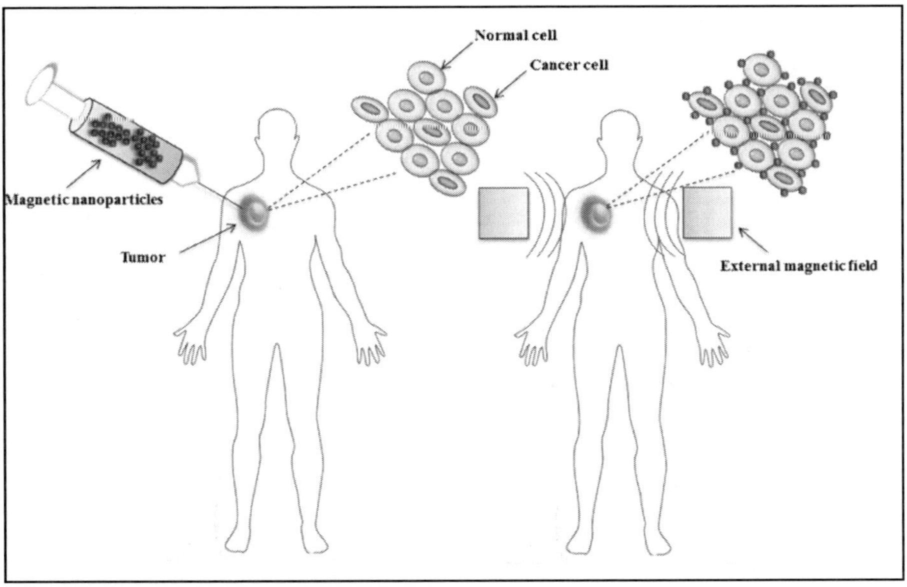

Figure 2. Magnetic hyperthermia treatment.

The second factor is the SAR which measures the efficiency of the heating power of a magnetic material and is described by the amount of energy converted into heat per unit of time and mass. Expressed in watts per gram (W/g) of material (Yi and Papaefthymiou, 2013), the SAR value depends mainly of the saturation magnetization of the MNP. On its turn, the saturation magnetization depends on the chemical composition, crystal structure, size, shape and surface modification of the particles. Further, the SAR value depends on the strength magnetic field (H) and frequency (f) of the applied magnetic field (Prasad, 2012). Many studies are being developed to improve the SAR of the MNPs by increasing the heating efficiency and reducing the amount of material that must be injected to treat a malignant tumor.

The third factor is the surface modification, where the MNPs are modified with components that are biocompatible and the modification has the further function of stabilizing the particles and prevent their agglomeration. Some of the compounds used to modify the surface of the nanoparticle are shown in Section 3. Some of the investigations that have been developed about the use of MNPs in magnetic hyperthermia are mentioned below.

Li et al. (Li et al., 2010b) synthesized MNPs of diameters ranging from 8 nm to 413 nm using a chemical co-precipitation and oxidation precipitation methods and investigated to what size of the MNP the highest heating efficiency in presence of an external magnetic field would be observed. Li et al. (Li et al., 2010b) placed 58 mg Fe/mL in an agar phantom and used magnetic field strengths of 120 Oe and 300 Oe, at a frequency of 100 kHz. The highest SAR values (52.8 W/g e 134.1 W/g) were obtained at 300 Oe for the sizes ranging from 8 nm to 24 nm, respectively. As for the temperature increase (ΔT) of the agar phantom due to the presence of the MNPs, the authors have founded that in 30 s the MNPs of 24 nm [8nm] produced a ΔT = 55 °C [25 °C]. As a rule, the MNPs exhibited had better heating efficiencies that the corresponding micrometric particles, a fact that has been attributed to a combination of effects Néel and Brown relaxations and hysteresis losses. Hence, MNPs with a diameter of 8 nm - 24 nm may be used as thermo-seeds for hyperthermia treatment.

Alphandéry et al. (Alphandéry et al., 2012) reported the use of chains of magnetosomes from magnetospirillum magneticum strain AMB-1 magnetotactic bacteria. The magnetosome chains are prokaryotic structures present in magnetotatic bacteria that contain in their inside magnetite crystals (15 to 20 crystals) that together act like a compass needle to orient the magnetotactic bacteria relative to the local direction of Earth´s magnetic field, helping them to move preferentially to microaerophile environments (Komeili et al., 2006). Magnetosome chains were more efficient for the inhibition of the proliferation of CA cells compared to individual ones, this response was attributed to the high cellular internalization and great stability in solution of the chains, which allowed an efficient heating. On the other hand, individual magnetosomes tend to agglomerate and so do not lead to an efficient inhibition of CA cell' growth. The main reason for the interest of using magnetosomes in magnetic hyperthermia for CA treatment is that the SAR value of biologically synthesized NPs are greater than those of chemically synthesized NPs.

Saavedra et al. (Martín-Saavedra et al., 2010) embedded NPs of maghemite (γ-Fe_2O_3) into a mesoporous silica-matrix to form magnetic microspheres (MMS). In order to evaluate their application in hyperthermia, they internalized the MMS in human tumor cells (A549, Saos-2 and HepG2 corresponding a lung, bone and liver cancers cells, respectively) and then found that as the MMS do not interfere with morphological features or metabolic activities of the cells, they present good biocompatibility. Saavedra et al. (Martín-Saavedra et al., 2010) used different MMS doses (48 mg to 160 mg) in a culture medium (1 mL) exposed to an alternating magnetic field (AMF) to observe the effect of heating of the MMS on the human tumor cells. It was also observed that increasing the amount of MMS increase the temperature of the cell culture medium; the temperature range varied from 44 °C to 52 °C, in a demonstration of the applicability of MMS in localized hyperthermia protocols.

Lin et al. (Lin et al., 2012) incorporated Fe_3O_4 MNPs into biodegradable cross-linked electrospun chitosan nanofiber for hyperthermia. The authors prepared magnetic nanofibers by two different methods, in the first one chitosan nanofibers were directly immersed in a solution of MNPs and in the other, chitosan nanofibers were submerged in a Fe^{2+}/Fe^{3+} solution, and then a chemical co-precipitation was made. They incubated Caco-2 (human colon adenocarcinoma) cells in a medium and then added the different magnetic nanofibers in the presence of a magnetic field, in both cases the temperature increased between 44 °C and 45 °C and remained constant. As a consequence of this heat generation the cell viability was decreased.

Hassan et al. (Hejase et al., 2012) synthesized manganese zinc ($Mn_xZn_{1-x}Fe_2O_4$) MNPs through co-precipitation and studied as the variation of the elemental composition of the MNPs would change their magnetic properties. Hassan et al. (Hejase et al., 2012) varied the Mn/Zn ratio by using Mn concentrations of $x=0.8$, 0.61, 0.5 and 0.2, and observed that the T_c of the NPs increased with the decreasing of Mn concentration (i.e., with a larger amount Zn in the sample). This behavior was attributed to the fact that the Zn atoms occupy tetrahedral sites in the lattice and displacing the Mn atoms from these sites; as the Zn atoms does not contribute to the net magnetic moment of the particles, a decrease of the T_c would take place. The corresponding T_c values were 205.4 °C, 128.5 °C, 103.0 °C and 53.3 °C for $x=0.8$, 0.61, 0.5 and 0.2, respectively.

Chalkidou et al. (Chalkidou et al., 2011) used Fe/MnO NPs on different human breast cancer cell to assess cytotoxicity, measure the intracellular uptake and evaluate the heating effect of NPs on the cancer cells. They obtained an outstanding cytotoxicity effect and a significant cell uptake efficiency (42-126 pg Fe/cell) with relatively high SAR values (100-500 W/g Fe) and fast thermal response (15 °C/10min).

Bai et al. (Bai et al., 2012) have reported a method of chemically depositing Fe_3O_4 NPs on graphene oxide nanosheets (Fe_3O_4 NPs-GON). Graphene is used due for its large surface area available to immobilize a high number of MNPs. To evaluate the Fe_3O_4 NPs-GON for hyperthermia application, the system was placed in a saline solution and then an alternating current (AC) magnetic field was applied during 500 s. After this time, the solution reached a temperature of 92.8 °C and so this compound may also be used thermo-seeds for hyperthermia treatment. In Table 2 we list some compounds that have been used for magnetic hyperthermia.

Table 2. MNPs used in magnetic hyperthermia

Magnetic compound	Modification of the MNP	Property evaluated	Ref.
γ-Fe_2O_3 and Fe_3O_4	-	ΔT=55 °C ΔT=25 °C SAR=134.1 W/g SAR=52.8 W/g	(Li et al., 2010b)
Fe_3O_4	Chitosan	Tc 53.7 °C Tc 53.1 °C T= 44-45 °C ΔT=23 °C	(Zhao et al., 2009) (Qu et al., 2010) (Lin et al., 2012) (Kim et al., 2006)
Fe_3O_4 magnetic fluid	Oleic acid and Tween 80	Tc 51 °C	(Wang et al., 2011)
γ-Fe_2O_3	Silica	Tc 44.0 °C	(Martín-Saavedra et al., 2010)
FeCo-Au	-	SAR=6.1 W/g	(Kline et al., 2009)
Fe_3O_4	Starch	ΔT=12 °C	(Kim et al., 2006)
$NiCo_2O_4$	mercaptopropionic acid	T=75 °C	(Kale et al., 2012)
Co-Ti ferrite	-	ΔT=15 °C	(Ichiyanagi et al., 2012)
$Mn_xZn_{1-x}Fe_2O_4$	-	Tc=56.3 °C	(Hejase et al., 2012)
Fe_3O_4	Lauric acid	SAR=16.5 W/g	(Chen et al., 2010)

Magnetic compound	Modification of the MNP	Property evaluated	Ref.
Cu-Ni	polyethylene glycol	Tc=43-46 °C	(Chatterjee et al., 2005)
Fe/MgO	-	SAR=100-500 W/g ΔT=15 °C	(Chalkidou et al., 2011)
Graphene oxide nanosheets-Fe_3O_4 NPs	-	T=92.8 °C	(Bai et al., 2012)
Cu-Ni	-	Tc=45 °C	(Ban et al., 2011)

4.3. Controlled Drug Release

Nowadays, increased attention is being dedicated to the use of MNPs in drug delivery systems (DDS) by exploiting their ability to reach a targeted area under guidance of an external magnetic field (Oh and Park, 2011). In fact, the applicability of MNPs in the development of DDS methods is related to a varied convenient factors for biotechnological applications such as biocompatibility, non-toxicity and non-immunogenicity (Ito et al., 2005; Oh and Park, 2011). However, in addition, for magnetic particles to be used as drug carriers it is required that they present a suitable surface coating that could avoid their recognition by the mononuclear phagocytic system (Liu et al., 2011).

Also, it is essential the use of protective agents to avoid aggregation of the MNPs. Silica has been extensively used as a protective layer due to its interesting characteristics for the development of targeting systems such as stability and low cytotoxicity, it is also possible to chemically modify its surface to produce more stable particles (Do Kim et al., 2007). Since biocompatibility and low toxicity are essential properties that new materials must possess to find use in biological applications, the addition of protective agents is a key step for the development of DDS (Oh and Park, 2011).

Recently, Anirudhan et al. (Anirudhan et al., 2013a) developed a hydrogel based on synthesized chitosan cross-linked β-cyclodextrin grafted silylated magnetic nanoparticle (CTSCD-g-SilylMNP) for delivery of Indomethacin (IND), a nonsteroidal anti-inflammatory drug used to reduce fever, pain and inflammation. A pH-dependent swelling property was observed for the CTSCD-g-SilylMNP system due to the chitosan crosslinking. In addition, the authors observed a swelling and diffusion controlled release for CTSCD-g-SilylMNP.

Other authors developed a magnetic N-benzyl-O-carboxymethylchitosan NPs loaded with IND for drug release (Debrassi et al., 2011). The corresponding MNPs were obtained by in situ precipitation methods and the nanoparticles, which contained a polymer cover, had an estimated drug loading efficiency between 60.8 % and 74.8 %. The kinetics of the drug release process for this system indicated a Fickian diffusion mechanism, and the magnetic properties of MNPs were not significantly affected by the incorporation of the drug (Debrassi et al., 2011).

Typically, magnetic core-shell systems consists of a magnetic core coated with a stabilizing shell to which targeting ligands are anchored. In addition, therapeutic agents can be embedded in the shell of the magnetic core-shell systems or chemically bonded on its surface. Naturally, the size, charge, hydrophobicity, shape, and orientation of the MNPs constituent materials are factors to be considered with regard to the overall NP physicochemical properties (Veiseh et al., 2010).

Wang et al. (Wang et al., 2013a) developed a very stable drug carrier by covalently coating magnetic nanoparticles (MNPs) with PEO-PPO-PEO block copolymer Pluronic P85 (MNPs-PPP). The MNPs-PPP, which presented a magnetite core with typical diameters of 15 nm and a 50 nm hydrodynamic radius, exhibited superparamagnetic behavior with a large saturation magnetization (57.1 emu/g) at room temperature. The covalent bonds between amino groups on the MNPs surface and the carboxyl groups of the modified P85 molecules ensure the stability of the MNPs-PPP. Also, the MNPs-PPP are temperature-responsive due to the presence of the PPP layer on their surface: variations in temperature cause conformational change of the copolymer and can improve both the load and the delivery efficiency of the particles.

Jain et al. (Jain et al., 2008) developed a novel oleic acid-coated iron-oxide and pluronic-stabilized MNPs with dual functional properties for drug delivery and magnetic resonance imaging (MRI). Drug incorporation efficiency of doxorubicin and paclitaxel (alone or in combination) in MNPs modified was in the 74-95 % range. In general, the drugs incorporated had marginal effects on physical (size or zeta potential) and magnetization properties of the MNPs. It is relevant to note that, when combined, drugs incorporated in MNPs demonstrated synergistic antiproliferative effects in MCF-7 breast cancer cells. The authors determined the circulation half-life ($t_{1/2}$) from the changes in the MRI signal intensity in carotid arteries in mice, and observed that the $T_{1/2}$ of the MNPs was longer than Feridex IV ($t_{1/2}$ = 31.2 vs 6.4 min) (Jain et al., 2008).

Arias et al. (Arias et al., 2008) developed DDS for anticancer drugs 5-Fluorouracil (5-FU, 5-fluoro-2,4-pyrimidinedione) and Ftorafur (Tegafur, 5-fluoro-1-(tetrahydro-2-furyl)uracil) for parenteral administration. 5-FU and Ftorafur have a broad spectrum of activity against solid tumors of gastrointestinal tract, breast, head and neck. Poly(butylcyanoacrylate) has been choice for the development of DDS due to its biodegradability characteristics and the possibility for drug transport and subsequently release. For both drugs, the release profile was found to be biphasic, since the drug adsorbed on the surface was released rather rapidly (requires only 1 h to be fully released) compared to the drug incorporated in the polymer matrix (requires between 10 and 20h).

Fang et al. (Fang et al., 2012) developed a system composed by superparamagnetics iron oxide nanoparticles (SPION), a biodegradable and pH-sensitive poly (beta-amino ester) (PBAE) copolymer and a chemotherapeutic agent doxorubicin (DOX). A sustained stability in cell culture media up to 7 days and a strong r2 relaxivity of 146 mM^{-1}•s^{-1} for magnetic resonance imaging was obtained. In vitro evaluation of NP-DOX efficacy using drug-resistant C6 glioma cells showed a 300 % increase in cellular internalization and 65 % reduction of half maximal inhibitory concentration (IC50) when compared to free DOX.

It is worthy to note that several formulations of SPIONs were approved for medical imaging and therapeutic applications, as follows: Feridex IV® for liver and spleen imaging (Bonnemain, 1998), Combidex® for lymph node metastases imaging (Harisinghani et al., 2003) and Ferumoxytol® for iron replacement therapy (Singh et al., 2008).

New methods for the synthesis of targeting DDS to specific sites have also been studied in detail (Ke et al., 2011). In addition, the extensively research in the development of new DDS is devoted to the possibility of increasing the drug efficiency during the treatment and reducing of the toxic effects on healthy cells and tissues (Sanna et al., 2011).

Majd et al. (Majd et al., 2013) engineered multifunctional nanomedicines for simultaneous imaging and therapy of cancer based on tamoxifen (TMX) loaded folic acid armed PEGylated MNPs to target the folate receptor positive cancer cells. Cytotoxicity analysis showed significant growth inhibition in MCF-7 cells treated with modified-MNPs.

To understand the factors that affect magnetic targeting of brain tumor lesions, Chertok et al. (Chertok et al., 2011) studied strategies for carotid administration of MNPs and the selective magnetic entrapment. Since the aggregation of nanoparticles in the afferent vasculature interfered with tumor targeting, the authors performed a simple modification of magnetic field topography to avoid the aggregation of NPs in the afferent vasculature. The carotid administration was found to present 1.8-fold increase in nanoparticle accumulation in glioma compared to the intravenous route at an applied magnetic field of 350 mT (Chertok et al., 2011).

4.4. MRI Contrast Enhancement

MRI is a noninvasive diagnostic technique used in medicine to investigate and detect different pathologies such as tumors, bleeding, injury, blood vessel diseases and others (Lee et al., 2010). This technique is based on physical phenomenon of nuclear magnetic resonance (NMR) (Hornak, 2013).

To obtain the MR image, the patient is subjected to a magnetic field produced by a magnet (usually the magnet is of 1.5 Tesla) which align the hydrogen protons (found in water molecules present in the tissues) in direction of the field, subsequently by applying a transverse radiofrequency pulses which causes a disturbance in the alignment of the protons (Berry and Curtis, 2003; Sun et al., 2008). When the frequency is suspend the protons are again influenced by the magnetic field and tries to realign with it along the longitudinal axis and return to their original state, this phenomenon is called relaxation (Sun et al., 2008). This relaxation phenomena occurs by two independent processes, longitudinal relaxation (T1) and transverse relaxation (T2). T1 is the time it takes the 63 % magnetization longitudinal in return to their alignment along the magnetic field and T2 is the time it takes for the transverse magnetization to decrease to 37 % of its initial value (Mornet et al., 2004). These relaxations can be monitored to generate a MR image. Each tissue type has its respective relaxation times causing differences of contrast in the image and allows us to distinguish between from one tissue to another (Saboktakin et al., 2009). To enhance the contrast in MR image contrast agents are used, which decrease T1 and T2 relaxation times by affecting the behavior of the protons in their proximity (Saboktakin et al., 2009). The contrast agents may be administered to the body by injection into the blood stream or orally. The increase of the proton relaxation rates determinate the efficiency of the contrast agent and is expressed by the r1 (1/T1) and r2 (1/T2) relaxivities (Henoumont et al., 2009). The value of r2/r1 ratio indicates whether will be generated a T1 or T2 weighted image. If the r2/r1 ratio is low, generates bright regions in the image (positive contrast) therefore the agent is considered T1. And if r2/r1 ratio is high are generate dark spots in the image (negative contrast) and the agent is considered T2 (Jedlovszky-Hajdú et al., 2012). In other words, a T2 contrast agent decreases the transverse relaxation time and a T1 decrease the longitudinal relaxation time.

Currently, the most common contrast agents used are T1 paramagnetic gadolinium (Gd) chelates, the problem of using these agents is that the Gd is toxic and furthermore Gd ions can replace some endogenous metal ions in the organism (Jedlovszky-Hajdú et al., 2012). Other T1 contrast agents that have been used in MRI are those of paramagnetic manganese (Bae et al., 2011). For this reason is being investigated new contrast agents, among the most promising is the superparamagnetic iron oxide (SPIO), which has some advantages in relation to common contrast agents, for example, do not retain remnant magnetization upon the removal of the magnetic field, have high magnetic properties, are biocompatible and non-toxic (Rosen et al., 2012). SPIO can produce a T2 relaxation effect, where the T2 shortening is due to that the MNPs create microscopic magnetic field gradients within the medium induces a dephasing of the proton magnetic moment (Schleich et al., 2013; Sun et al., 2008).

Today, a wide variety of particles have been synthetized with different types of coating for MRI such as polyesters (Lee et al., 2004a), chitosan (Lee et al., 2010), carboxylic compounds (Jedlovszky-Hajdú et al., 2012), polyethylene glycol (Kumagai et al., 2007), pluronic copolymer (Jain et al., 2009), dextran (Erten et al., 2010) and starch (Saboktakin et al., 2009). The most of these systems have multifunctional applications, because they can be used to diagnose (MRI) and treat (hyperthermia and drug delivery) diseases (Schleich et al., 2013). Someone of the systems that have been developed are mentioned below.

Jedlovsky-Hajdú et al. (Jedlovszky-Hajdú et al., 2012) prepared a magnetic fluid make of Fe_3O_4 NPs modified with different carboxylic compounds (citric acid, polyacrylic acid (PAA) and sodium oleate (NAaOA)). To evaluate the use of these MNPs as contrast agents in MRI, were performed relaxation measurements at different field strengths (0.47 T, 1.5 T and 9.4 T) and the results were compared with ferucarbotran (a well-known superparamagnetic contrast agent). They found that both r1 and r2 relaxivities depend of the magnetic field strength, and their highest values of relaxivities were obtained to r2 at 1.5 T. The high r2/r1 ratios of the different magnetic fluid show that these materials can be good negative contrast agents in MRI.

Yang et al. (Yang et al., 2012) synthetized superparamagnetic Fe_3O_4 NPs modified with silica and subsequently functionalized with thymidine. This system was used as a nanosensor for Hg (II) and was used MR image to calculate the relaxation times. Hg is one of the metals that cause major health problems can damage the nervous and endocrine system, so it is important to detect this element in the environment and human body. The interest of use MRI sensors is that can be used to detect Hg even in turbid system. They found that by adding Hg (II) to the solution containing the functionalized NPs decrease the transverse relaxation time (T2). This decrease was attributed to the interaction between the thymidine/mercury causing the formation of an aggregate, which disturbs the spins of water protons, leading to an enhancement of r2.

Yallapu et al. (Yallapu et al., 2011) reported the use of superparamagnetic Fe_3O_4 NPs coated by β-cyclodextrin or by pluronic polymer for hyperthermia, magnetic resonance imaging (MRI) and drug delivery applications. In the case of *in vitro* MRI, they proved pure and modified NPs in ovarian cancer cells and were evaluated the relaxation times of the different system. The results showed that the system more efficient was Fe_3O_4 coated with pluronic polymer, due to that show the largest decrease in the transverse relaxation time (T2).

Choi et al. (Choi et al., 2004) synthetized Fe_2O_3 NPs coated with dextran and were tethered N-hydroxysuccinimide-folate and fluorescence isothiocyanate, these NPs were used as contrast agents for *in vivo* MRI. For *in vivo* test were used mice with subcutaneous tumor xenografts grown from implanted KB cells (human nasopharyngeal epidermal carcinoma cell). MR image were carried before and after the NPs incorporation, the results showed that NPs incorporation decreases the average intensity of the image in 38% (on the tumor area) compared with the image before treatment, indicating that these NPs can be used as contrast agents T2 in MRI.

Bae et al. (Bae et al., 2011) reported the use of hollow manganese oxide nanoparticles (HMON) modified with 3,4-dihydroxy-L-phenylalanine (DOPA) for immobilization of polyethylenimine (PEI) and the subsequent conjugation of Herceptin (HER). HMON-PEI-HER NPs were used to cancer treatment by to use siRNA and simultaneously diagnose by MRI. The siRNA is a therapeutic agent that can help treatment various diseases because it can interfere with the expression of specific genes. The Herceptin is a monoclonal antibody that specifically bind the human epidermal growth factor receptor 2 (HER2); where an amplification or over-expression of HER2 has shown an increase in the development of certain types of cancer. To bind the herceptin to the HER2 the NPs release the siRNA suppressing the growth of cancer cells. To evaluate the use of HMON-PEI-HER NPs as contrast agents were performed *in vitro* tests using breast cancer cells (SK-BR-3 and MCF-7). The MR images were obtained with a 3.0 T clinical MRI scanner. The results indicate that to incorporate the NPs to cancer cell culture take place an increase in the bright signal in relation to not-treated cells, indicating that these particles can be used to enhance the positive T1 contrast.

4.5. Biosensors

MNPs can be used to form matrices for immobilization of biomolecules, such as proteins, enzymes, and other bioactive agents in analytical biochemistry, medicine, and biotechnology (Kouassi and Irudayaraj, 2006). In addition, these biomolecules immobilized on magnetic nanoparticles have been applied as biosensors (Grancharov et al., 2005).

In general, biosensors are devices that can detect analytes by combining a biological recognition element with a physicochemical responsive component. The mainly elements of a biosensor are the biological recognition element, the transducer or the detector and the reader device (Plata et al., 2010).

Biosensors are promising alternatives for identification of harmful pathogens (Pal and Alocilja, 2009). A new antibody/gold nanoparticle /magnetic nanoparticle nanocomposites (anti-body/AuNP/MNPs) for the detection of the food borne pathogen *Staphylococcus aureus* (S. aureus) in milk, with bases on colorimetric detection (Sung et al., 2013), whose detection limits for S. aureus in buffer and the milk sample were 1.5×10^3 CFU and 1.5×10^5 CFU, respectively. Its sensing mechanism also had a higher specificity for S. aureus relative to other pathogens such as *Escherichia coli*, *Listeria monocytogenes*, and *Salmonella enterica* (Sung et al., 2013).

Pal and Alocilja (Pal and Alocilja, 2009) developed an electrically active polyaniline coated magnetic (EAPM) nanoparticle-based biosensor for the detection of *Bacillus anthracis endospores* in contaminated food samples, such as lettuce, ground beef and whole milk samples. The detection mechanism of that biosensor depend on how the spores captured on the biosensor surface would capillary flow during the direct-charge transfer across the EAPM nanoparticles, and its detection limit correspond to concentrations as low as 4.2×10^2 spores/mL from the samples.

A wireless magnetoelastic-sensing device was developed for detection of *Escherichia coli* O157:H7 (E. coli), by use of chitosan-modified magnetic

Fe_3O_4 nanoparticles (CMNPs) as signal amplifying tags (Lin et al., 2010). The E. coli attached CMNPs were magnetically bound to the surface of the magnetoelastic sensor. The sensor showed a linear response in the range of 10 cells mL^{-1} to 3.7×10^8 cells mL^{-1}, with a lower detection limit of 10 cells mL^{-1}.

Also the direct electrocatalysis and electrochemistry of heme proteins entrapped in carbon-coated nickel magnetic nanoparticle-chitosan-dimethylformamide (CNN-CS-DMF) composite films were investigated (Wang et al., 2013b). The addition of CNN significantly enhanced not only the electron transfer of the heme proteins but also their electrocatalytic activity toward the reduction of H_2O_2. The CNN-CS-DMF system exhibited a lower detection limit and improved sensitivity for detecting H_2O_2. Recently, Fe_3O_4 mesoporous spheres modified carbon ionic liquid electrode was applied to the development of electrochemical hemoglobin biosensor (Sun et al., 2013).

An interdigitated array microelectrode (IDAM) coupled with magnetic nanoparticle-antibody conjugates (MNAC) was developed and evaluated for rapid and specific detection of E. coli O157:H7 in ground beef samples (Varshney and Li, 2007). The lowest detection limits of this biosensor for detection of E. coli O157:H7 in pure culture and ground beef samples were 7.4×10^4 and 8.0×10^5 CFU mL^{-1}, respectively. The sensitivity of the impedance biosensor was improved by 35 % by use of a magnetic field to concentrate the cells attached to MNAC in the active layer of IDAM above the surface of electrodes with the help of a magnetic field.

Recently, Cristescu et al. (Cristescu et al., 2012) reported thin film deposition of nanostructured Fe_3O_4/oleic acid/ceftriaxone and Fe_3O_4/oleic acid/cefepime nanoparticles (core/shell/adsorption-shell) fabricated by the matrix assisted pulsed laser evaporation (MAPLE) technique onto inert substrates. The microbial viability and the microbial adherence to the glass coverslip nanoparticle film was studied using Gram-negative and Gram-positive bacterial strains. In addition, the antibiotic charged NPs exhibited a strong inhibitory effect on the viability as well as on the adherence to the cellular monolayer.

An enzyme biosensor based on Fe_3O_4 nanoparticles-chitosan nanocomposites has been developed for the detection of phenolic compounds (Wang et al., 2008), with a linear response range for concentrations between 8.3×10^{-8} to 7.0×10^{-5} $mol.L^{-1}$, and a detection limit of 2.5×10^{-8} $mol.L^{-1}$. Both the large surface area of Fe_3O_4 nanoparticles and the porous morphology of chitosan contributes to a high loading of enzyme. The entrapped molecules in the Fe_3O_4-polymer systems retained the bioactivity (Wang et al., 2008).

Layer-by-layer assembly technique has also been used with magnetic core-shell Au-Fe_3O_4@SiO_2 nanocomposite to fabricate a bienzyme glucose biosensor based on glucose oxidase and horseradish peroxidase (Chen et al., 2011). In this case, the biosensor showed a linear response in the 0.05-1.0 mM and 1.0-8.0 mM ranges, while the detection limit was estimated as 0.01 mM.

Baby and Ramaprabhu developed a new type of amperometric glucose biosensor based on silicon dioxide coated magnetic nanoparticle decorated multiwalled carbon nanotubes (Fe_3O_4@SiO_2/MWNTs) on a glassy carbon electrode (GCE) (Baby and Ramaprabhu, 2010). This glucose biosensor exhibited a linear response from 1 μM to 30 mM with an excellent lower detection limit of 800 nM.

Recently, polydopamine (PDA)-laccase (Lac)-Fe_3O_4 magnetic polymeric bionanocomposites (MPBNCs) were prepared through one-pot Lac-catalyzed oxidation of dopamine in an aqueous suspension containing Lac, magnetic Fe_3O_4 nanoparticles and dopamine (Li et al., 2012). A PDA-Lac-Fe_3O_4/Au electrode was prepared by facile and efficient magnetic separation/ immobilization of the MPBNCs on a magnetic Au electrode for biosensing of hydroquinone, giving a sensitivity of 374 μA $mmol^{-1}$ L cm^{-2}, and a detection limit of 30 nmol L^{-1}.

Optical techniques have also been applied in SPR bioassays based on amplification techniques using core-shell Fe_3O_4@Au magnetic nanoparticles (MNPs) (Liang et al., 2012). The authors developed a novel SPR sensor, with a detection limit of 0.65 ng mL^{-1}, based on a sandwich immunoassay to detect α-fetoprotein (AFP) by immobilizing a primary AFP antibody (Ab1) on the surface of a 3-mercapto-1-propanesulfonate/chitosan-ferrocene/Au NP (MPS/CS-Fc/Au NP) film employing Fe_3O_4@Au-AFP secondary antibody conjugates (Fe_3O_4@Au–Ab2) as the amplification reagent.

ACKNOWLEDGMENTS

The authors are grateful for support from the Rede de Nanobiotecnologia /CAPES and INCT_IF (Instituto Nacional de Ciência e Tecnologia para Inovação Farmacêutica), CNPq (grant 310305/2012-8 and 310361/2012-5), CAPES and FACEPE.

References

Alibeigi, S., and Vaezi, M. R. (2008). Phase Transformation of Iron Oxide Nanoparticles by Varying the Molar Ratio of Fe2+:Fe3+. *Chemical Engineering & Technology 31*, 1591-1596.

Alphandéry, E., Guyot, F., and Chebbi, I. (2012). Preparation of chains of magnetosomes, isolated from Magnetospirillum magneticum strain AMB-1 magnetotactic bacteria, yielding efficient treatment of tumors using magnetic hyperthermia. *International Journal of Pharmaceutics 434*, 444-452.

Anirudhan, T. S., Dilu, D., and Sandeep, S. (2013a). Synthesis and characterisation of chitosan crosslinked-beta-cyclodextrin grafted silylated magnetic nanoparticles for controlled release of Indomethacin. *J. Magn. Magn. Mater. 343*, 149-156.

Anirudhan, T. S., Dilu, D., and Sandeep, S. (2013b). Synthesis and characterisation of chitosan crosslinked-β-cyclodextrin grafted silylated magnetic nanoparticles for controlled release of Indomethacin. *Journal of Magnetism and Magnetic Materials 343*, 149-156.

Arias, J. L., Linares-Molinero, F., Gallardo, V., and Delgado, A. V. (2008). Study of carbonyl iron/poly(butylcyanoacrylate) (core/shell) particles as anticancer drug delivery systems - Loading and release properties. *Eur. J. Pharm. Sci. 33*, 252-261.

Ashtari, K., Khajeh, K., Fasihi, J., Ashtari, P., Ramazani, A., and Vali, H. (2012). Silica-encapsulated magnetic nanoparticles: Enzyme immobilization and cytotoxic study. *International Journal of Biological Macromolecules 50*, 1063-1069.

Ashtari, P., He, X., Wang, K., and Gong, P. (2005). An efficient method for recovery of target ssDNA based on amino-modified silica-coated magnetic nanoparticles. *Talanta 67*, 548-554.

Baby, T. T., and Ramaprabhu, S. (2010). SiO2 coated Fe3O4 magnetic nanoparticle dispersed multiwalled carbon nanotubes based amperometric glucose biosensor. *Talanta 80*, 2016-2022.

Bae, K. H., Lee, K., Kim, C., and Park, T. G. (2011). Surface functionalized hollow manganese oxide nanoparticles for cancer targeted siRNA delivery and magnetic resonance imaging. *Biomaterials 32*, 176-184.

Bai, L.-Z., Zhao, D.-L., Xu, Y., Zhang, J.-M., Gao, Y.-L., Zhao, L.-Y., and Tang, J.-T. (2012). Inductive heating property of graphene oxide–Fe3O4 nanoparticles hybrid in an AC magnetic field for localized hyperthermia. *Materials Letters 68*, 399-401.

Ban, I., Stergar, J., Drofenik, M., Ferk, G., and Makovec, D. (2011). Synthesis of copper–nickel nanoparticles prepared by mechanical milling for use in magnetic hyperthermia. *J. Magn. Magn. Mater. 323*, 2254-2258.

Bele, M., Hribar, G., Čampelj, S., Makovec, D., Gaberc-Porekar, V., Zorko, M., Gaberšček, M., Jamnik, J., and Venturini, P. (2008). Zinc-decorated silica-coated magnetic nanoparticles for protein binding and controlled release. *Journal of Chromatography* B *867*, 160-164.

Berry, C. C., and Curtis, A. S. G. (2003). Functionalization of magnetic nanoparticles for applications in biomedicine. *Journal of Physics D: Applied Physics 36*, R198-R206.

Bhaumik, M., Maity, A., Srinivasu, V. V., and Onyango, M. S. (2011). Enhanced removal of Cr(VI) from aqueous solution using polypyrrole/Fe3O4 magnetic nanocomposite. *Journal of Hazardous Materials 190*, 381-390.

Bonnemain, B. (1998). Superparamagnetic agents in magnetic resonance imaging: Physicochemical characteristics and clinical applications - A review. *J. Drug Target 6*, 167-174.

Cao, M., Li, Z., Wang, J., Ge, W., Yue, T., Li, R., Colvin, V. L., and Yu, W. W. (2012). Food related applications of magnetic iron oxide nanoparticles: Enzyme immobilization, protein purification, and food analysis. *Trends in Food Science & Technology 27*, 47-56.

Chalkidou, A., Simeonidis, K., Angelakeris, M., Samaras, T., Martinez-Boubeta, C., Balcells, L., Papazisis, K., Dendrinou-Samara, C., and Kalogirou, O. (2011). In vitro application of Fe/MgO nanoparticles as magnetically mediated hyperthermia agents for cancer treatment. *J. Magn. Magn. Mater. 323*, 775-780.

Charles P. Poole, J., and Owens, F. J. (2003). Introduction to nanotechnology (Wiley-Interscience).

Chatterjee, J., Bettge, M., Haik, Y., and Jen Chen, C. (2005). Synthesis and characterization of polymer encapsulated Cu–Ni magnetic nanoparticles for hyperthermia applications. *J. Magn. Magn. Mater. 293*, 303-309.

Chen, S., Chiang, C.-l., and Hsieh, S. (2010). Simulating physiological conditions to evaluate nanoparticles for magnetic fluid hyperthermia (MFH) therapy applications. *Journal of Magnetism and Magnetic Materials 322*, 247-252.

Chen, X. J., Zhu, J. W., Chen, Z. X., Xu, C. B., Wang, Y., and Yao, C. (2011). A novel bienzyme glucose biosensor based on three-layer Au-Fe3O4@SiO2 magnetic nanocomposite. *Sensor Actuat B-Chem. 159*, 220-228.

Chertok, B., David, A. E., and Yang, V. C. (2011). Brain tumor targeting of magnetic nanoparticles for potential drug delivery: Effect of administration route and magnetic field topography. *J. Control Release 155*, 393–399.

Cherukuri, P., Glazer, E. S., and Curley, S. A. (2010). Targeted hyperthermia using metal nanoparticles. *Advanced Drug Delivery Reviews 62*, 339-345.

Chiang, C.-L., Sung, C.-S., Wu, T.-F., Chen, C.-Y., and Hsu, C.-Y. (2005). Application of superparamagnetic nanoparticles in purification of plasmid DNA from bacterial cells. *Journal of Chromatography* B *822*, 54-60.

Choi, H., Choi, S.R., Zhou, R., Kung, H.F., and Chen, I.W. (2004). Iron oxide nanoparticles as magnetic resonance contrast agent for tumor imaging via folate receptor-targeted delivery1. Academic Radiology *11*, 996-1004.

Cristescu, R., Popescu, C., Socol, G., Iordache, I., Mihailescu, I. N., Mihaiescu, D. E., Grumezescu, A. M., Balan, A., Stamatin, I., Chifiriuc, C., et al. (2012). Magnetic core/shell nanoparticle thin films deposited by MAPLE: Investigation by chemical, morphological and in vitro biological assays. *Appl. Surf. Sci. 258*, 9250-9255.

Debrassi, A., Burger, C., Rodrigues, C. A., Nedelko, N., Slawska-Waniewska, A., Dluzewski, P., Sobczak, K., and Greneche, J. M. (2011). Synthesis, characterization and in vitro drug release of magnetic N-benzyl-O-carboxymethylchitosan nanoparticles loaded with indomethacin. *Acta Biomater. 7*, 3078-3085.

DeNardo, S. J., DeNardo, G. L., Natarajan, A., Miers, L. A., Foreman, A. R., Gruettner, C., Adamson, G. N., and Ivkov, R. (2007). Thermal Dosimetry Predictive of Efficacy of 111In-ChL6 Nanoparticle AMF–Induced Thermoablative Therapy for Human Breast Cancer in Mice. *Journal of Nuclear Medicine 48*, 437-444.

Deng, J., He, C., Peng, Y., Wang, J., Long, X., Li, P., and Chan, A. S. C. (2003). Magnetic and conductive Fe3O4–polyaniline nanoparticles with core–shell structure. *Synthetic Metals 139*, 295-301.

Do Kim, K., Min, S. S., Choa, Y. H., and Kim, H. T. (2007). Formation and surface modification of Fe3O4 nanoparticles by co-precipitation and sol-gel method. *J. Ind. Eng. Chem. 13*, 1137-1141.

Erten, A., Wrasidlo, W., Scadeng, M., Esener, S., Hoffman, R. M., Bouvet, M., and Makale, M. (2010). Magnetic resonance and fluorescence imaging of doxorubicin-loaded nanoparticles using a novel in vivo model. *Nanomedicine: nanotechnology, biology, and medicine 6*, 797-807.

Fang, C., Kievit, F. M., Veiseh, O., Stephen, Z. R., Wang, T. Z., Lee, D. H., Ellenbogen, R. G., and Zhang, M. Q. (2012). Fabrication of magnetic nanoparticles with controllable drug loading and release through a simple assembly approach. *J. Control Release 162*, 233-241.

Frey, N. A., Peng, S., Cheng, K., and Sun, S. (2009). Magnetic nanoparticles: synthesis, functionalization, and applications in bioimaging and magnetic energy storage. *Chemical Society Reviews 38*, 2532-2542.

Georgelin, T., Moreau, B., Bar, N., Villemin, D., Cabuil, V., and Horner, O. (2008). Functionalization of gamma-Fe2O3 nanoparticles through the grafting of an organophosphorous ligand. *Sensor Actuat. B-Chem. 134*, 451-454.

Ghaemy, M., and Naseri, M. (2012). Synthesis of chitosan networks: Swelling, drug release, and magnetically assisted BSA separation using Fe3O4 nanoparticles. *Carbohydrate Polymers 90*, 1265-1272.

Grancharov, S. G., Zeng, H., Sun, S. H., Wang, S. X., O'Brien, S., Murray, C. B., Kirtley, J. R., and Held, G. A. (2005). Bio-functionalization of monodisperse magnetic nanoparticles and their use as biomolecular labels in a magnetic tunnel junction based sensor. *J. Phys. Chem*. B *109*, 13030-13035.

Harisinghani, M. G., Barentsz, J., Hahn, P. F., Deserno, W. M., Tabatabaei, S., van de Kaa, C. H., de la Rosette, J., and Weissleder, R. (2003). Noninvasive detection of clinically occult lymph-node metastases in prostate cancer. *New Engl. J. Med. 348*, 2491-U2495.

Hee Kim, E., Sook Lee, H., Kook Kwak, B., and Kim, B.-K. (2005). Synthesis of ferrofluid with magnetic nanoparticles by sonochemical method for MRI contrast agent. *Journal of Magnetism and Magnetic Materials 289*, 328-330.

Hejase, H., Hayek, S. S., Qadri, S., and Haik, Y. (2012). MnZnFe nanoparticles for self-controlled magnetic hyperthermia. *Journal of Magnetism and Magnetic Materials 324*, 3620-3628.

Henoumont, C., Laurent, S., and Vander Elst, L. (2009). How to perform accurate and reliable measurements of longitudinal and transverse relaxation times of MRI contrast media in aqueous solutions. *Contrast Media & Molecular Imaging 4*, 312-321.

Hornak, J. P. (2013). The Basics of MRI (J. P Hornak).

Ichiyanagi, Y., Shigeoka, D., Hiroki, T., Mashino, T., Kimura, S., Tomitaka, A., Ueda, K., and Takemura, Y. (2012). Study on increase in temperature of Co–Ti ferrite nanoparticles for magnetic hyperthermia treatment. *Thermochimica Acta 532*, 123-126.

Institute, N. C. (2011). Hyperthermia in Cancer Treatment.

Ito, A., and Honda, H. (2007). Magnetic Nanoparticles for Tissue Engineering. In Nanotechnologies for the Life Sciences (Wiley-VCH Verlag GmbH & Co. KGaA).

Ito, A., Shinkai, M., Honda, H., and Kobayashi, T. (2005). Medical application of functionalized magnetic nanoparticles. *J. Biosci. Bioeng. 100*, 1-11.

Jain, T. K., Foy, S. P., Erokwu, B., Dimitrijevic, S., Flask, C. A., and Labhasetwar, V. (2009). Magnetic resonance imaging of multifunctional pluronic stabilized iron-oxide nanoparticles in tumor-bearing mice. *Biomaterials 30*, 6748-6756.

Jain, T. K., Richey, J., Strand, M., Leslie-Pelecky, D. L., Flask, C. A., and Labhasetwar, V. (2008). Magnetic nanoparticles with dual functional properties: Drug delivery and magnetic resonance imaging. *Biomaterials 29*, 4012-4021.

Jedlovszky-Hajdú, A., Tombácz, E., Bányai, I., Babos, M., and Palkó, A. (2012). Carboxylated magnetic nanoparticles as MRI contrast agents: Relaxation measurements at different field strengths. *Journal of Magnetism and Magnetic Materials 324*, 3173-3180.

Jiang, C., Liu, R., Shen, X., Zhu, L., and Song, F. (2011). Ni0.5Zn0.5Fe2O4 nanoparticles and their magnetic properties and adsorption of bovine serum albumin. *Powder Technology 211*, 90-94.

Johannsen, M., Gnevekow, U., (2010). Magnetic nanoparticle hyperthermia for prostate cancer. *International Journal of Hyperthermia 26*, 790-795.

Kale, S. N., Jadhav, A. D., Verma, S., Koppikar, S. J., Kaul-Ghanekar, R., Dhole, S. D., and Ogale, S. B. (2012). Characterization of biocompatible NiCo2O4 nanoparticles for applications in hyperthermia and drug delivery. *Nanomedicine: Nanotechnology, Biology and Medicine 8*, 452-459.

Käppler, T. E., Hickstein, B., Peuker, U. A., and Posten, C. (2008). Characterization of magnetic ion-exchange composites for protein separation from biosuspensions. *Journal of Bioscience and Bioengineering 105*, 579-585.

Ke, C. J., Su, T. Y., Chen, H. L., Liu, H. L., Chiang, W. L., Chu, P. C., Xia, Y. N., and Sung, H. W. (2011). Smart Multifunctional Hollow Microspheres for the Quick Release of Drugs in Intracellular Lysosomal Compartments. *Angew Chem. Int. Edit. 50*, 8086-8089.

Kim, D. H., Lee, S. H., Im, K. H., Kim, K. N., Kim, K. M., Shim, I. B., Lee, M. H., and Lee, Y. K. (2006). Surface-modified magnetite nanoparticles for hyperthermia: Preparation, characterization, and cytotoxicity studies. Current Applied Physics *6, Supplement 1*, e242-e246.

Kline, T. L., Xu, Y.-H., Jing, Y., and Wang, J.-P. (2009). Biocompatible high-moment FeCo-Au magnetic nanoparticles for magnetic hyperthermia treatment optimization. *Journal of Magnetism and Magnetic Materials 321*, 1525-1528.

Komeili, A., Li, Z., Newman, D. K., and Jensen, G. J. (2006). Magnetosomes Are Cell Membrane Invaginations Organized by the Actin-Like Protein MamK. *Science 311*, 242-245.

Kouassi, G. K., and Irudayaraj, J. (2006). Magnetic and gold-coated magnetic nanoparticles as a DNA sensor. *Anal. Chem. 78*, 3234-3241.

Kumagai, M., Imai, Y., Nakamura, T., Yamasaki, Y., Sekino, M., Ueno, S., Hanaoka, K., Kikuchi, K., Nagano, T., Kaneko, E., et al. (2007). Iron hydroxide nanoparticles coated with poly(ethylene glycol)-poly(aspartic acid) block copolymer as novel magnetic resonance contrast agents for in vivo cancer imaging. *Colloids and Surfaces B: Biointerfaces 56*, 174-181.

Lai, B.-H., Yeh, C.-C., and Chen, D.-H. (2012). Surface modification of iron oxide nanoparticles with polyarginine as a highly positively charged magnetic nano-adsorbent for fast and effective recovery of acid proteins. *Process Biochemistry 47*, 799-805.

Laurent, S., Forge, D., Port, M., Roch, A., Robic, C., Vander Elst, L., and Muller, R. N. (2008). Magnetic Iron Oxide Nanoparticles: Synthesis, Stabilization, Vectorization, Physicochemical Characterizations, and Biological Applications. *Chemical Reviews 108*, 2064-2110.

Lee, P.-W., Hsu, S.-H., Wang, J.-J., Tsai, J.-S., Lin, K.-J., Wey, S.-P., Chen, F.-R., Lai, C.-H., Yen, T.-C., and Sung, H.-W. (2010). The characteristics, biodistribution, magnetic resonance imaging and biodegradability of superparamagnetic core–shell nanoparticles. *Biomaterials 31*, 1316-1324.

Lee, S.-J., Jeong, J.-R., Shin, S.-C., Kim, J.-C., Chang, Y.-H., Chang, Y.-M., and Kim, J.-D. (2004a). Nanoparticles of magnetic ferric oxides encapsulated with poly(D,L latide-co-glycolide) and their applications to magnetic resonance imaging contrast agent. *Journal of Magnetism and Magnetic Materials 272–276, Part 3*, 2432-2433.

Lee, S.-J., Jeong, J.-R., Shin, S.-C., Kim, J.-C., and Kim, J.-D. (2004b). Synthesis and characterization of superparamagnetic maghemite nanoparticles prepared by coprecipitation technique. *J. Magn. Magn. Mater. 282*, 147-150.

Li, G.-y., Zhou, Z.-d., Li, Y.-j., Huang, K.-l., and Zhong, M. (2010a). Surface functionalization of chitosan-coated magnetic nanoparticles for covalent immobilization of yeast alcohol dehydrogenase from Saccharomyces cerevisiae. *Journal of Magnetism and Magnetic Materials 322*, 3862-3868.

Li, X., Zhang, J., and Gu, H. (2011). Adsorption and Desorption Behaviors of DNA with Magnetic Mesoporous Silica Nanoparticles. *Langmuir 27*, 6099-6106.

Li, Y. Y., Qin, C., Chen, C., Fu, Y. C., Ma, M., and Xie, Q. J. (2012). Highly sensitive phenolic biosensor based on magnetic polydopamine-laccase-Fe3O4 bionanocomposite. *Sensor Actuat. B-Chem. 168*, 46-53.

Li, Z., Kawashita, M., Araki, N., Mitsumori, M., Hiraoka, M., and Doi, M. (2010b). Magnetite nanoparticles with high heating efficiencies for application in the hyperthermia of cancer. *Materials Science and Engineering*: C *30*, 990-996.

Liang, R. P., Yao, G. H., Fan, L. X., and Qiu, J. D. (2012). Magnetic Fe3O4@Au composite-enhanced surface plasmon resonance for ultrasensitive detection of magnetic nanoparticle-enriched alpha-fetoprotein. *Anal. Chim. Acta 737*, 22-28.

Lin, H. L., Lu, Q. Z., Ge, S. T., Cai, Q. Y., and Grimes, C. A. (2010). Detection of pathogen Escherichia coli O157:H7 with a wireless magnetoelastic-sensing device amplified by using chitosan-modified magnetic Fe3O4 nanoparticles. *Sensor Actuat. B-Chem. 147*, 343-349.

Lin, T.-C., Lin, F.-H., and Lin, J.-C. (2012). In vitro feasibility study of the use of a magnetic electrospun chitosan nanofiber composite for hyperthermia treatment of tumor cells. *Acta Biomaterialia 8*, 2704-2711.

Liu, G., Wang, Z. Y., Lu, J., Xia, C. C., Gao, F. B., Gong, Q. Y., Song, B., Zhao, X. N., Shuai, X. T., Chen, X. Y., et al. (2011). Low molecular weight alkyl-polycation wrapped magnetite nanoparticle clusters as MRI probes for stem cell labeling and in vivo imaging. *Biomaterials 32*, 528-537.

Liu, X., Xing, J., Guan, Y., Shan, G., and Liu, H. (2004a). Synthesis of amino-silane modified superparamagnetic silica supports and their use for protein immobilization. *Colloids and Surfaces A: Physicochemical and Engineering Aspects 238*, 127-131.

Liu, Z. L., Liu, Y. J., Yao, K. L., Ding, Z. H., Tao, J., and Wang, X. (2002). Synthesis and Magnetic Properties of Fe3O4 Nanoparticles. *Journal of Materials Synthesis and Processing 10*, 83-87.

Liu, Z. L., Wang, H. B., Lu, Q. H., Du, G. H., Peng, L., Du, Y. Q., Zhang, S. M., and Yao, K. L. (2004b). Synthesis and characterization of ultrafine well-dispersed magnetic nanoparticles. *J. Magn. Magn. Mater. 283*, 258-262.

Lu, A.-H., Salabas, E. L., and Schüth, F. (2007). Magnetic Nanoparticles: Synthesis, Protection, Functionalization, and Application. *Angewandte Chemie International Edition 46*, 1222-1244.

Ma, M., Zhang, Y., Yu, W., Shen, H. Y., Zhang, H. Q., and Gu, N. (2003). Preparation and characterization of magnetite nanoparticles coated by amino silane. *Colloid Surface A 212*, 219-226.

Majd, M. H., Asgari, D., Barar, J., Valizadeh, H., Kafil, V., Abadpour, A., Moumivand, E., Mojarrad, J. S., Rashidi, M. R., Coukos, G., et al. (2013). Tamoxifen loaded folic acid armed PEGylated magnetic nanoparticles for targeted imaging and therapy of cancer. *Colloid Surface B 106*, 117-125.

Martín-Saavedra, F. M., Ruíz-Hernández, E., Boré, A., Arcos, D., Vallet-Regí, M., and Vilaboa, N. (2010). Magnetic mesoporous silica spheres for hyperthermia therapy. *Acta Biomater. 6*, 4522-4531.

Meng, J., Shi, C., Wei, B., Yu, W., Deng, C., and Zhang, X. (2011). Preparation of Fe3O4@C@PANI magnetic microspheres for the extraction and analysis of phenolic compounds in water samples by gas chromatography–mass spectrometry. *Journal of Chromatography A 1218*, 2841-2847.

Mornet, S., Vasseur, S., Grasset, F., and Duguet, E. (2004). Magnetic nanoparticle design for medical diagnosis and therapy. *Journal of Materials Chemistry 14*, 2161-2175.

Oh, J. K., and Park, J. M. (2011). Iron oxide-based superparamagnetic polymeric nanomaterials: Design, preparation, and biomedical application. *Prog. Polym. Sci. 36*, 168-189.

Organization, W. H. (January 2013). *Cancer*.

Pal, S., and Alocilja, E. C. (2009). Electrically active polyaniline coated magnetic (EAPM) nanoparticle as novel transducer in biosensor for detection of Bacillus anthracis spores in food samples. *Biosens Bioelectron 24*, 1437-1444.

Pham, T. A., Kumar, N. A., and Jeong, Y. T. (2010a). Facile preparation of boronic acid functionalized Fe-core/Au-shell magnetic nanoparticles for covalent immobilization of adenosine. *Colloids and Surfaces A: Physicochemical and Engineering Aspects 370*, 95-101.

Pham, T. A., Kumar, N. A., and Jeong, Y. T. (2010b). Facile preparation of boronic acid functionalized Fe-core/Au-shell magnetic nanoparticles for covalent immobilization of adenosine. *Colloids and Surfaces A: Physicochem. Eng. Aspects 370*, 95-101.

Pham, T. A., Kumar, N. A., and Jeong, Y. T. (2010c). Facile preparation of boronic acid functionalized Fe-core/Au-shell magnetic nanoparticles for covalent immobilization of adenosine. *Colloid Surface A 370*, 95-101.

Plata, M. R., Contento, A. M., and Rios, A. (2010). State-of-the-Art of (Bio)Chemical Sensor Developments in Analytical Spanish Groups. *Sensors-Basel 10*, 2511-2576.

Prasad, P. N. (2012). Introduction to nanomedicine and nanobioengineering (WILEY).

Qu, J., Liu, G., Wang, Y., and Hong, R. (2010). Preparation of Fe3O4–chitosan nanoparticles used for hyperthermia. *Advanced Powder Technology 21*, 461-467.

Qun Zhao, L. W., Rui Cheng, Leidong Mao, Robert D. Arnold, Elizabeth W. Howerth, Zhuo G. Chen, and Simon Platt (2012). Magnetic nanoparticle-based hyperthermia for head and neck cancer in mouse models. *Theranostics* 2, *2*, 113–121.

Roca, A. G., Marco, J. F., Morales, M. D. P., and Serna, C. J. (2007). Effect of Nature and Particle Size on Properties of Uniform Magnetite and Maghemite Nanoparticles. *The Journal of Physical Chemistry* C *111*, 18577-18584.

Rosen, J. E., Chan, L., Shieh, D.-B., and Gu, F. X. (2012). Iron oxide nanoparticles for targeted cancer imaging and diagnostics. *Nanomedicine: Nanotechnology, Biology and Medicine 8*, 275-290.

Saboktakin, M. R., Maharramov, A., and Ramazanov, M. A. (2009). Synthesis and characterization of superparamagnetic nanoparticles coated with carboxymethyl starch (CMS) for magnetic resonance imaging technique. *Carbohydrate Polymers 78*, 292-295.

Sanna, V., Pintus, G., Bandiera, P., Anedda, R., Punzoni, S., Sanna, B., Migaleddu, V., Uzzau, S., and Sechi, M. (2011). Development of Polymeric Microbubbles Targeted to Prostate-Specific Membrane Antigen as Prototype of Novel Ultrasound Contrast Agents. Mol Pharmaceut *8*, 748-757.

Scarberry, K. E., Dickerson, E. B., Zhang, Z. J., Benigno, B. B., and McDonald, J. F. (2010). Selective removal of ovarian cancer cells from human ascites fluid using magnetic nanoparticles. *Nanomedicine: Nanotechnology, Biology and Medicine 6*, 399-408.

Schleich, N., Sibret, P., Danhier, P., Ucakar, B., Laurent, S., Muller, R. N., Jérôme, C., Gallez, B., Préat, V., and Danhier, F. (2013). Dual anticancer drug/superparamagnetic iron oxide-loaded PLGA-based nanoparticles for cancer therapy and magnetic resonance imaging. *International Journal of Pharmaceutics 447*, 94-101.

Shen, X. C., Fang, X. Z., Zhou, Y. H., and Liang, H. (2004). Synthesis and characterization of 3-aminopropyltriethoxysilane-modified superparamagnetic magnetite nanoparticles. *Chem. Lett. 33*, 1468-1469.

Singh, A., Patel, T., Hertel, J., Bernardo, M., Kausz, A., and Brenner, L. (2008). Safety of Ferumoxytol in Patients With Anemia and CKD. *Am. J. Kidney Dis. 52*, 907-915.

Spaldin, N. A. (2003). Magnetic Materials, Fundamentals and Applications, second edition edn (Cambridge).

Srirajaskanthan, R., and Preedy, V. R. (2012). Nanomedicine and Cancer (Science Publishers).

Sulek, F., Drofenik, M., Habulin, M., and Knez, K. (2010). Surface functionalization of silica-coated magnetic nanoparticles for covalent attachment of cholesterol oxidase. *J. Magn. Magn. Mater. 322*, 179–185.

Sun, C., Lee, J. S. H., and Zhang, M. (2008). Magnetic nanoparticles in MR imaging and drug delivery. *Advanced Drug Delivery Reviews 60*, 1252-1265.

Sun, W., Sun, Z. L., Zhang, L. Q., Qi, X. W., Li, G. J., Wu, J., and Wang, M. (2013). Application of Fe3O4 mesoporous sphere modified carbon ionic liquid electrode as electrochemical hemoglobin biosensor. *Colloid Surface B 101*, 177-182.

Sung, Y. J., Suk, H. J., Sung, H. Y., Li, T., Poo, H., and Kim, M. G. (2013). Novel antibody/gold nanoparticle/magnetic nanoparticle nanocomposites for immunomagnetic separation and rapid colorimetric detection of Staphylococcus aureus in milk. *Biosens. Bioelectron. 43*, 432-439.

Tan, M. C. (2009). Nanostructured Materials for Biomedical Applications. (Transworld Research Network), p. 120.

Tartaj, P., Morales, M. D. P., Veintemillas-Verdaguer, S., González-Carreño, T., and Serna, C. J. (2003). The preparation of magnetic nanoparticles for applications in biomedicine. *Journal of Physics D: Applied Physics 36*, R182.

Tartaj, P., Morales, M. P., González-Carreño, T., Veintemillas-Verdaguer, S., and Serna, C. J. (2005). Advances in magnetic nanoparticles for biotechnology applications. *J. Magn. Magn. Mater. 290–291, Part 1*, 28-34.

Varshney, M., and Li, Y. B. (2007). Interdigitated array microelectrode based impedance biosensor coupled with magnetic nanoparticle-antibody conjugates for detection of Escherichia coli O157: H7 in food samples. *Biosens. Bioelectron. 22*, 2408-2414.

Veiseh, O., Gunn, J. W., and Zhang, M. Q. (2010). Design and fabrication of magnetic nanoparticles for targeted drug delivery and imaging. *Adv. Drug Deliver Rev. 62*, 284-304.

Wang, N., Guan, Y. P., Yang, L. R., Jia, L. W., Wei, X. T., Liu, H. Z., and Guo, C. (2013a). Magnetic nanoparticles (MNPs) covalently coated by PEO-PPO-PEO block copolymer for drug delivery. *J. Colloid Interf. Sci. 395*, 50-57.

Wang, S. F., Tan, Y. M., Zhao, D. M., and Liu, G. D. (2008). Amperometric tyrosinase biosensor based on Fe(3)O(4) nanoparticles-chitosan nanocomposite. *Biosens. Bioelectron. 23*, 1781-1787.

Wang, T., Wang, L., Tu, J., Xiong, H., and Wang, S. (2013b). Direct electrochemistry and electrocatalysis of heme proteins immobilised in carbon-coated nickel magnetic nanoparticle–chitosan–dimethylformamide composite films in room-temperature ionic liquids. Bioelectrochemistry 94 (2013) 94–99. *Bioelectrochemistry 94*, 94–99.

Wang, Y. M., Cao, X., Liu, G. H., Hong, R. Y., Chen, Y. M., Chen, X. F., Li, H. Z., Xu, B., and Wei, D. G. (2011). Synthesis of Fe3O4 magnetic fluid used for magnetic resonance imaging and hyperthermia. *J. Magn. Magn. Mater. 323*, 2953-2959.

Wust, P., Hildebrandt, B., Sreenivasa, G., Rau, B., Gellermann, J., Riess, H., Felix, R., and Schlag, P. M. (2002). Hyperthermia in combined treatment of cancer. *The Lancet Oncology 3*, 487-497.

Yallapu, M. M., Othman, S. F., Curtis, E. T., Gupta, B. K., Jaggi, M., and Chauhan, S. C. (2011). Multi-functional magnetic nanoparticles for magnetic resonance imaging and cancer therapy. *Biomaterials 32*, 1890-1905.

Yang, H., Tian, Z., Wang, J., and Yang, S. (2012). A magnetic resonance imaging nanosensor for Hg (II) based on thymidine-functionalized supermagnetic iron oxide nanoparticles. *Sensors and Actuators B: Chemical 161*, 429-433.

Yi, D. K., and Papaefthymiou, G. C. (2013). Nanobiomaterials, Development and applications. (CRC Press).

Zhao, D.-L., Wang, X.-X., Zeng, X.-W., Xia, Q.-S., and Tang, J.-T. (2009). Preparation and inductive heating property of Fe3O4–chitosan composite nanoparticles in an AC magnetic field for localized hyperthermia. *Journal of Alloys and Compounds 477*, 739-743.

Zhou, Z. H. A. J. W. (2001). Synthesis of Fe3O4 nanoparticles from emulsions. *Journal of Materials Chemistry*, 1704-1709.

In: Magnetic Nanoparticles
Editor: Nora P. Sabbas
ISBN: 978-1-63117-434-6

Chapter 2

CARBON–COATED FUNCTIONALIZED MAGNETIC NANOPARTICLES FOR BIOMEDICAL APPLICATIONS

***Mercedes Arana*[1], *Paula Bercoff*[1], *Silvia Jacobo*[2] *and Marcela Rodríguez*[3]**

[1]Facultad de Matemática, Astronomía y Física,
Universidad Nacional de Córdoba. IFEG, Conicet. Córdoba, Argentina
[2]LAFMACEL, Facultad de Ingeniería, Universidad de Buenos Aires.
INTECIN, Conicet. Buenos Aires, Argentina
[3]Facultad de Ciencias Químicas, Universidad Nacional de Córdoba.
INFIQC, Conicet. Córdoba, Argentina

ABSTRACT

Carbon–coated magnetite nanoparticles (NPs) were synthetized by the mechanochemical method with hematite as precursor and amorphous carbon as inorganic reductor. After 18 hours of milling in an inert atmosphere, a nanocomposite material of magnetite and carbon was obtained.

Structural and magnetic properties of the NPs were investigated by X-ray diffraction (XRD), scanning electron microscopy (SEM), transmission electron microscopy (TEM), energy dispersive X-ray spectroscopy (EDS) and vibrating sample magnetometry. XRD patterns, refined with the Rietveld method, show that magnetite is present in samples milled from 6 hours onward and that after milling for 18 hours and annealing in Ar, the sample contains a single crystalline phase. Magnetization curves for samples with different milling times show saturation magnetization values that range from 34.1 emu/g after 1 h to 78.0 emu/g after 18 h. Coercive fields are about 500 Oe for all samples. TEM studies reveal that the samples are made of amorphous carbon clusters with magnetite NPs of 20 nm.

The obtained NPs, associated to electrochemical transducers, show an improved enhancement of the charge transfer for redox processes involving different bioanalytes. Thus, these NPs offer unique properties as a catalyst in biosensing strategies for the electrochemical detection of high-impact markers and the development of theranostics smart-devices for biomedical applications.

Index Terms—Carbon, coatings, magnetic nanoparticles, hydrogen peroxide, biosensor

I. Introduction

The physical and biological sciences share a common interest in nanoscale, since integrating biology and nanomaterials may improve different fields of science and technology. Nanomaterials and nanotechnology are widely applied in biomedicine, especially in the areas of biomedical diagnosis, drugs and implants [1]. The applications of biomedical nanotechnology outside the body include biosensors and biochips and inside the body include targeted drugs delivery, implantation of insulin pumps and gene therapy [2].

Nanoscience and nanotechnology comprise different biomedical applications for magnetic nanomaterials. Most of the biomedical applications of magnetic nanomaterials are based on their specific characteristics.

Though magnetic nanoparticles with different compositions, sizes and shapes have been developed for biomedical applications, the most frequently used magnetic materials are maghemite (γ-Fe_2O_3) and magnetite (Fe_3O_4) [3–9].

Nowadays, there is a rising demand for biosensors with high sensitivity and reliability, fast response and excellent selectivity. Biosensors have found versatile applications in the field of environmental control, hazardous materials detection, pharmaceutics and clinical diagnostics. A common biosensor works as a biospecific surface interacting with a particular analyte, and generating detectable signals, such as electrochemical, optical, piezoelectrical and thermal responses. The enzyme glucose oxidase (GOx) has been the most widely used biorecognition element in glucose biosensors [10]. GOx catalyzes the oxidation of glucose to gluconolactone and this, in turn, reduces the oxygen to hydrogen peroxide. The main challenge when designing electrochemical glucose biosensors is the development of strategies that allow an important decrease in the high overvoltage required for the oxidation and reduction of hydrogen peroxide (H_2O_2) at carbon electrodes in order to improve H_2O_2 quantification [11].

Comba et al [11–13] reported the advantages of using carbon paste electrodes modified with magnetite and iron nanoparticles in glucose biosensors. These authors found that sensor sensitivity is related to magnetite content. However, high NPs contents promote a highly resistive behavior due to the low–conductive nature of magnetite. These considerations led us to infer that magnetite-carbon interface could play an important role in the biosensor response. We decided to prepare carbon–coated magnetite nanoparticles in order to explore this idea.

Mechanical milling is recognized as an effective way of producing solid–state chemical reactions at low temperatures. This is a well–known process, with the first publication dating back to some decades ago [14], that induces chemical reactions in powder precursors during the collisions with the grinding media [15]. Mechanical milling can be classified as mechanical grinding, mechanical alloying, and mechanochemical synthesis in relation to the precursors, the milling parameters and the structural and chemical changes that occur during the milling. Mechanical grinding in a high–energy ball mill favors the formation of fine (nanosized) particles, while the chemical composition of precursors changes as a result of mechanically induced solid state reaction [15, 16].

Mechanochemical reactions are characterized by a large negative free energy change at room temperature and therefore they are thermodynamically favorable at room temperature [17, 18]. The preparation of soft ferrites by mechanosynthesis involves chemical precursors such as mixtures of oxides, chlorides and/or metals that react either during milling or during subsequent heat treatment, to form a new product or composite powder. In particular, it is possible to produce an 80 – 95% of spinel phase from a stoichiometric mixture of simple oxides as precursors, in a high–energy ball mill (in dry or wet media), while the single ferritic phase is usually obtained only with a posterior thermal annealing [19].

The applications of mechanochemistry include exchange reactions, reduction/oxidation reactions, decomposition of compounds, and phase transformations. Let M be a pure metal, R a reductant and MO the corresponding metal oxide reduced by R. Thus, the exchange reactions in mechanochemistry can be represented by an equation of the type $MO + R \rightarrow M + RO$.

In this work we report the preparation of carbon–coated magnetite NPs by mechanosynthesis, and their structural and magnetic characterization (part of which was previously reported in [20]). These NPs were included within carbon paste electrodes in order to explore their analytical performance towards hydrogen peroxide reduction and to evaluate their response as a glucose biosensor.

II. Experimental

Magnetite (Fe_3O_4) NPs were synthesized by the mechanochemical method in a *Fritsch Pulverisette 7* high–energy ball mill. The structural characterization of the samples was performed by X ray diffraction (XRD) with a Phillip X'Pert diffractometer, with Cu–K_α radiation, and Rietveld refinement using *FullProf Suite* software. The magnetic properties were determined with a Lake Shore 7300 vibrating sample magnetometer (VSM) at room temperature and with a maximum applied field of 1200 kA/m (1.5 kOe). The size distribution and the morphology of samples were investigated with a field emission scanning electron microscope (FE–SEM) *SIGMA, Zeiss* and with a high resolution transmission electron microscope (HR–TEM) *Philips CM 200* that is equipped with a LaB6 emission filament which operates at 200 keV.

The electrochemical measurements were performed with a TEQ_04 potentiostat. The electrodes were inserted into the cell (BAS, Model MF–1084) through holes in its Teflon cover. A platinum wire and Ag/AgCl, 3 M KCl (BAS, Model RE–5B) were used as counter and reference electrodes, respectively. All potentials are referred to the latter. A magnetic stirrer provided the convective transport during the amperometric measurements. The electrochemical characterization of magnetite NPs was performed using a composite carbon paste electrode (CPE). The CPE was prepared in a regular way by mechanically mixing graphite powder (70.0% w/w) and mineral oil (30.0% w/w) in an agate mortar for 30 min. CPEs containing magnetite NPs (CPE–NPs) and commercial magnetite microparticles (100–500 nm) (CPE–MPs) were prepared in a similar way, mixing first the magnetite NPs or commercial magnetite MPs with mineral oil for 3 min, followed by the incorporation of the graphite powder and mixing for additional 30 min. In the case of the enzymatic electrode, GOx and magnetite NPs were first mixed with the mineral oil for 10 min before incorporating the graphite powder and then mixed for additional 30 min. A portion of the resulting paste was packed firmly into the cavity (3 mm diameter) of a Teflon tube (3 mm diameter). The electric contact was established through a stainless steel screw. A new surface was obtained by smoothing the electrode onto a weighing paper before starting every new experiment. Cyclic voltammetric experiments were performed in 0.050 M buffer phosphate solution pH 7.40 as supporting electrolyte. The amperometric measurements were conducted in a stirred 0.050 M buffer phosphate solution pH 7.40 by applying the desired working potential and allowing the transient current to decay to a steady–state prior to the addition of the analyte and subsequent current monitoring. All the electrochemical measurements were performed at room temperature.

III. SYNTHESIS AND STRUCTURAL CHARACTERIZATION

High-energy Ball Milling

Hematite (Fe_2O_3) and activated carbon were used as precursors in a 4:1 mass ratio according to the following reaction:

$$3Fe_2O_3 + 10C \rightarrow 2Fe_3O_4 + CO + 9C \quad (1)$$

As reaction (1) indicates, extra C is introduced in the vials so that a carbon excess remains in the sample even after the hematite reduction process has finished, covering the formed magnetite particles. An inert atmosphere had to be used in the milling vials to avoid secondary oxidation processes.

The milling was performed at 700 rpm, with a ball/powder mass ratio of 35, in stainless steel vials with WC balls and Ar atmosphere. Every synthesis parameter was chosen after several tests and previous studies.

The samples that result from the milling are labeled CHt where *t* is the milling time in hours and takes the values 1, 2, 3, 6, 12 and 18.

Structural Characterization and Thermal Treatments

X ray diffraction patterns of hematite used as precursor and those of the resulting powders from the milling are shown in Figure 1.

From Figure 1, it is evident that the expected chemical reaction (1) starts at $t = 6$ h, but hematite is still present in the powder even after milling for 18 hours, indicating that the reaction is not complete. Activated carbon is amorphous; therefore, no reflections arising from this phase are expected with this technique.

In order to get rid of the remaining hematite in the samples and avoid further milling, the samples were annealed in Ar for 2 h at 500 °C. This temperature was carefully selected to prevent a significant particle growth as well as C evaporation [21] while promoting hematite reduction.

The resulting powders after annealing were labeled CHtT, being *t* the milling time in hours. X ray diffraction patterns for these samples are shown in Figure 2. Sample CH18T shows only magnetite reflections.

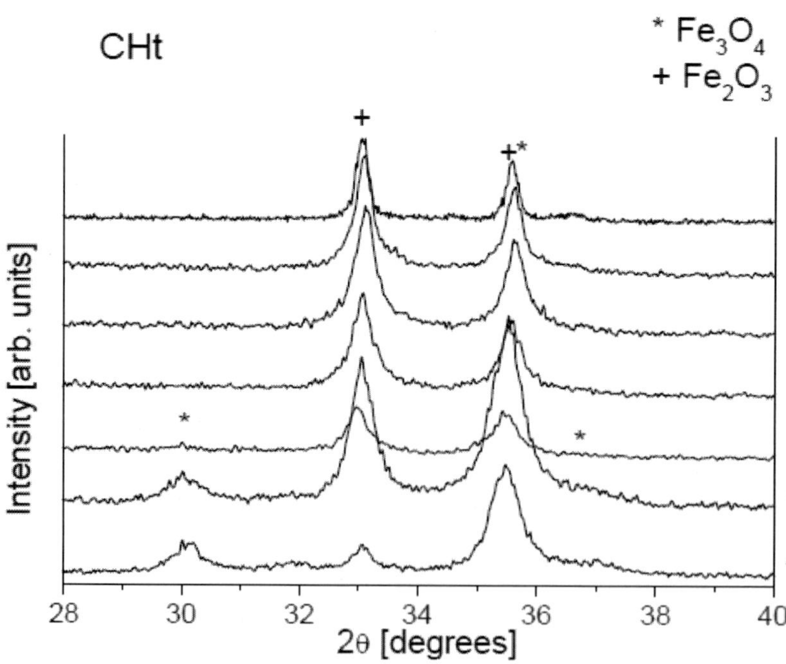

Figure 1. X ray diffraction patterns for samples milled from 1 to 18 h and precursor mixture (*t*=0).

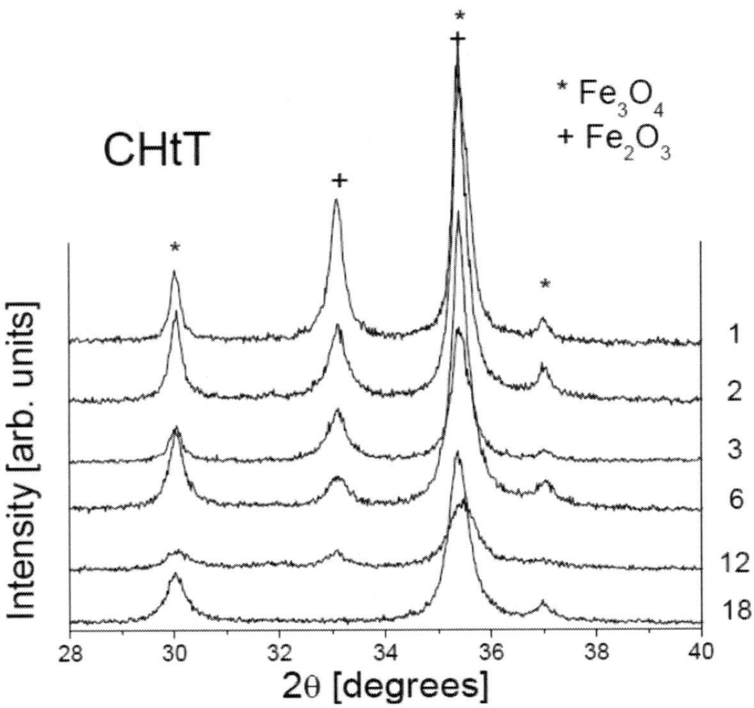

Figure 2. X ray diffraction patterns for samples milled from 1 to 18 h and annealed for 2 h at 500°C, in Ar atmosphere.

Rietveld Analysis

X ray diffraction patterns were refined with the Rietveld method. Figure 3 shows the fitting of the corresponding spectra of samples CH1T, CH3T and CH18T.

The conventional Rietveld *R*–factors (Chi^2, GoF and R_{Bragg} for each crystalline phase) for XRD refinements of samples CHtT, for all the milling times are presented in Table I, while the structural parameters values of magnetite for: cell parameter *a*, cell volume *V*, density ρ , crystallite size $D_{crystal}$ and phase content of each sample *P* obtained from Rietveld refinement data are shown in Table II.

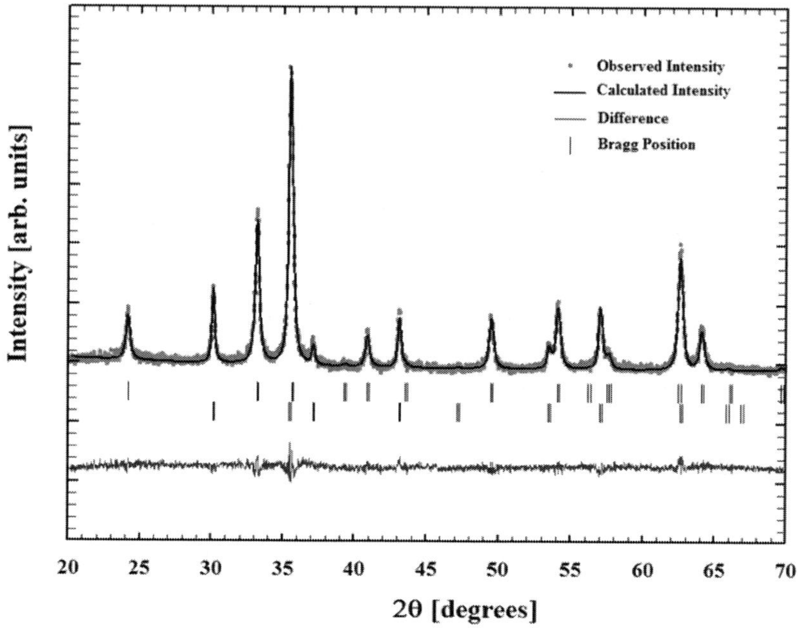

(a)

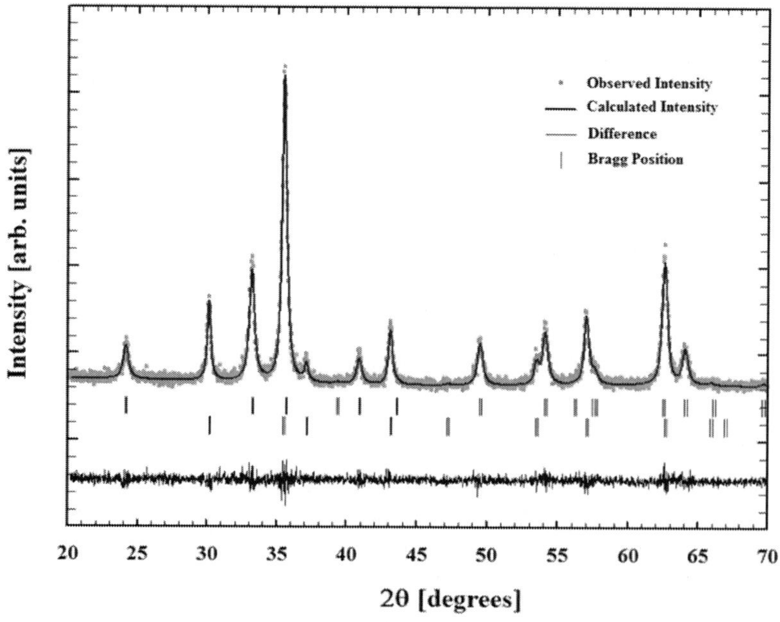

(b)

Figure 3. (Continued).

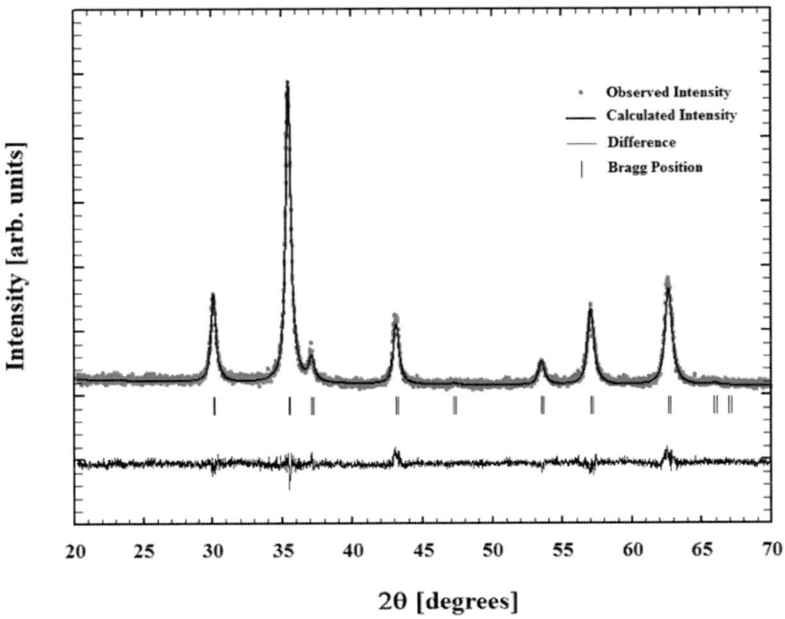

(c)

Figure 3. Rietveld refinements of the XRD pattern of samples CH1T (a), CH3T (b) and CH18T (c). Only magnetite reflections (vertical lines) are present in the pattern of sample CH18T. The dots are experimental data, the dark line is the theoretical pattern and the line below is the difference between experimental and theoretical data; vertical lines indicate the Bragg reflections.

Table I. Conventional Rietveld R-factors for samples CHtT

			R_{Bragg}	
Sample	Chi^2	GoF	Fe_2O_3	Fe_3O_4
CH1T	1.26	11.708	6.78	4.45
CH3T	1.13	4.511	4.33	2.73
CH6T	1.22	9.860	7.10	4.10
CH12T	1.55	24.326	26.0	10.60
CH18T	1.14	6.361	-	4.77

Even when the Rietveld method is quite accurate for determining crystallographic parameters and phases, it has proven to be rather tricky when an amorphous material is present in the sample. So, even when obtaining values of Chi^2 around 1 in every case, the overall quality parameter is high (indicating not a good overall fit). In order to take the amorphous fraction into account, a different approach to the refinement must be addressed. However, the parameters obtained for the crystalline phases (hematite and magnetite, in this case) are reliable.

The cell parameter of magnetite slightly changes with the milling time. This change can be attributed to the strains which are generated in the synthesis process due to the impacts of the mill balls with the powder. Besides, the cell volume is affected with the milling time in a similar way.

It is clear from Table II that the interplanar distance of magnetite disminishes when the milling time increases from 1 to 18 h, taking values from 0.4720 to 0.4716 Å, respectively. The density of ferritic phase remains almost constant, with changes of the order of the accuracy of the calculation method and with values ranging from 5.21 to 5.23 g/cm^3.

Table II. Rietveld refinement parameters for samples CHtT

Sample	a [±0.0001Å]	d [±0.0002Å]	V [±0.001Å^3]	ρ [±0.01g/cm^3]	$D_{crystal}$ [±1nm]	P [%] Fe_2O_3	P [%] Fe_3O_4
CH1T	8.3879	0.4720	590.1388	5.212	43	56.6	43.4
CH3T	8.3895	0.4721	590.4822	5.213	28	51.0	48.8
CH6T	8.3872	0.4719	589.9957	5.216	27	20.1	79.9
CH12T	8.3803	0.4716	588.5320	5.225	11	20.3	79.7
CH18T	8.3813	0.4716	588.7602	5.224	20	0.0	100.0

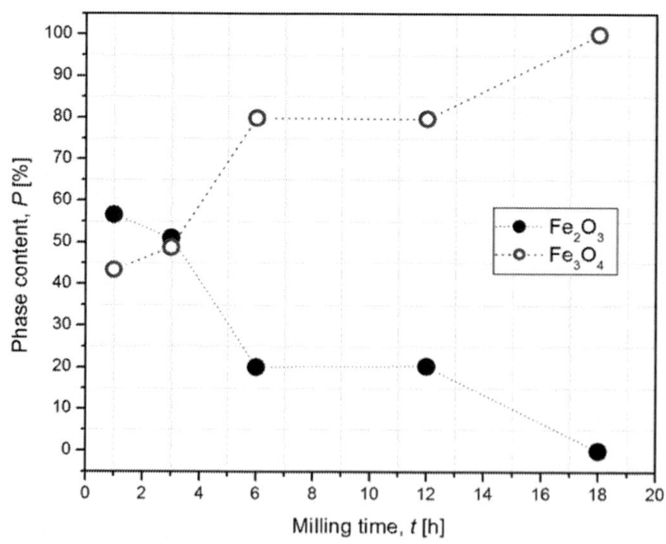

Figure 4. Hematite and magnetite content in the samples, as a function of milling time.

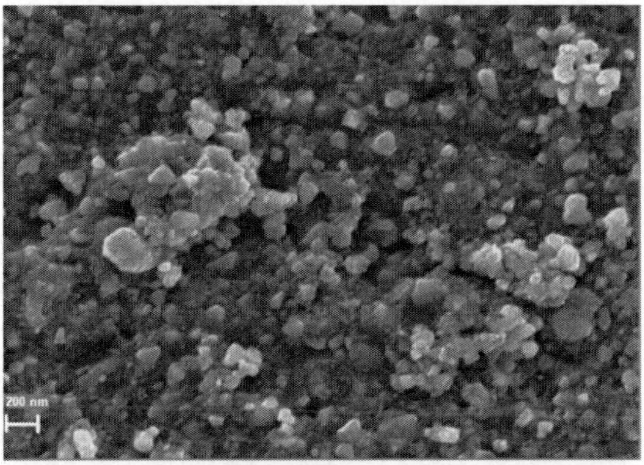

a)

b)

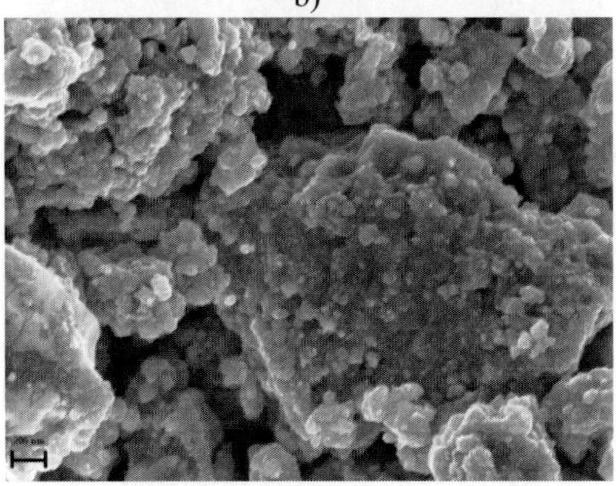

c)

Figure 5. (Continued).

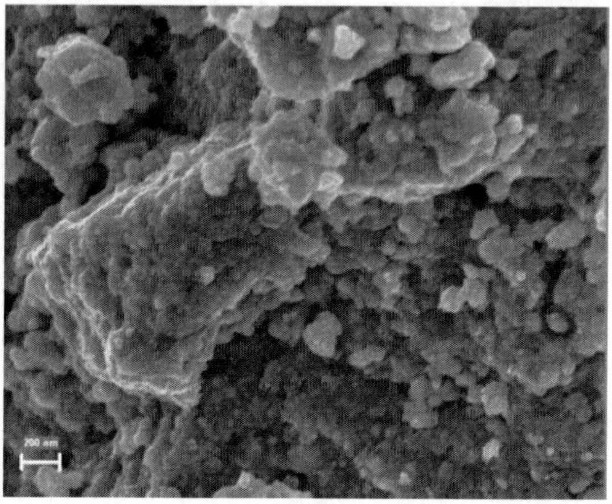

d)

Figure 5. SEM images for samples CHtT, for *t*=1 (a), 2(b), 3(c) and 6h (d).

Crystal size decreases from 43 to 11 nm for magnetite in samples milled from 1 to 12 h, and there is a small increase after 18 h of milling. This increase can be explained with the disappearance of hematite, which tends to prevent crystal growth [22].

The phase content P, as a function of milling time is presented in Figure 4. The reduction of hematite to magnetite as a function of milling time is evident, and is complete after 18 hours.

SEM and TEM Characterization

Scanning electron micrographs (SEM) show that that the particles are nanosized (see Figure 5) and appear in the powders as single particles and clusters.

Figure 6 shows secondary electron (a) and a backscattered electron (b) images of sample CH18T. In backscattered electron images, elements with a higher atomic number Z appear brighter than elements with lower Z. The comparison of figures 6 a and b indicates that nanometric crystals of ferrite are uniformly inmersed in carbon. This was further confirmed by transmission electron microscopy (TEM), which was also used to investigate the samples. In order to do so, the powders were dispersed in ethyl alcohol. Afterwards, small drops were deposited on ultrathin carbon support films (of the order of 4 nm thickness) which were then deposited on Cu grids and loaded in the TEM holder. High resolution transmission electron microscopy (HRTEM) images show that sample CH18 is composed of nanosized particles, surrounded by carbon (see Figure 7), accordingly with SEM results. In the images presented in Figure 7, both hematite and magnetite can be identified with the different d-spacings of the crystalline planes, as indicated by the arrows. The size distribution is Gaussian–shaped. The particle size (D_P) distribution in the inset shows that the average particle size from TEM (D_{PT}) is 11 nm.

a)

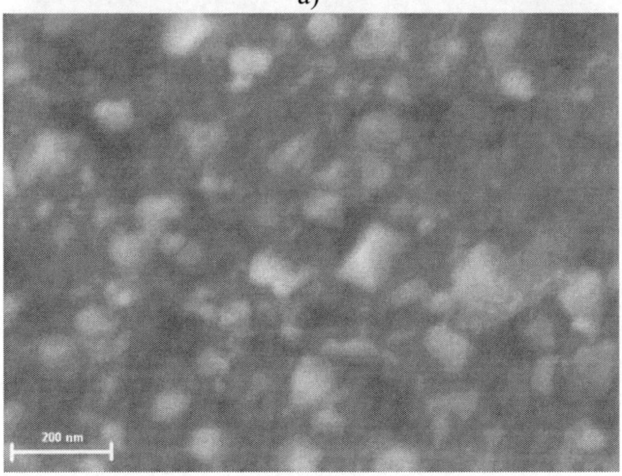

b)

Figure 6. SEM micrographs of sample CH18T. a) Secondary electrons image. b) Backscattered electrons image.

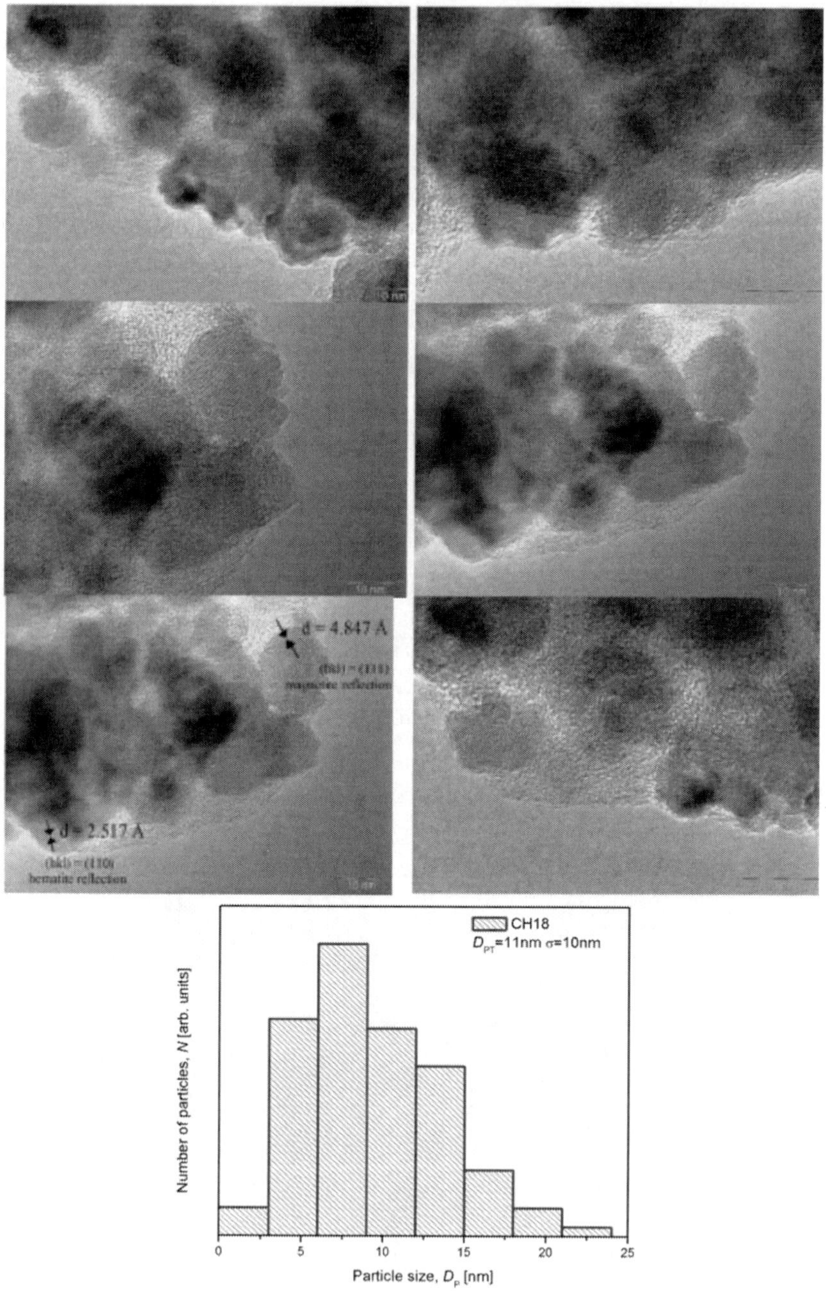

Figure 7. HRTEM images of sample CH18. The inset shows the Gaussian particle size distribution.

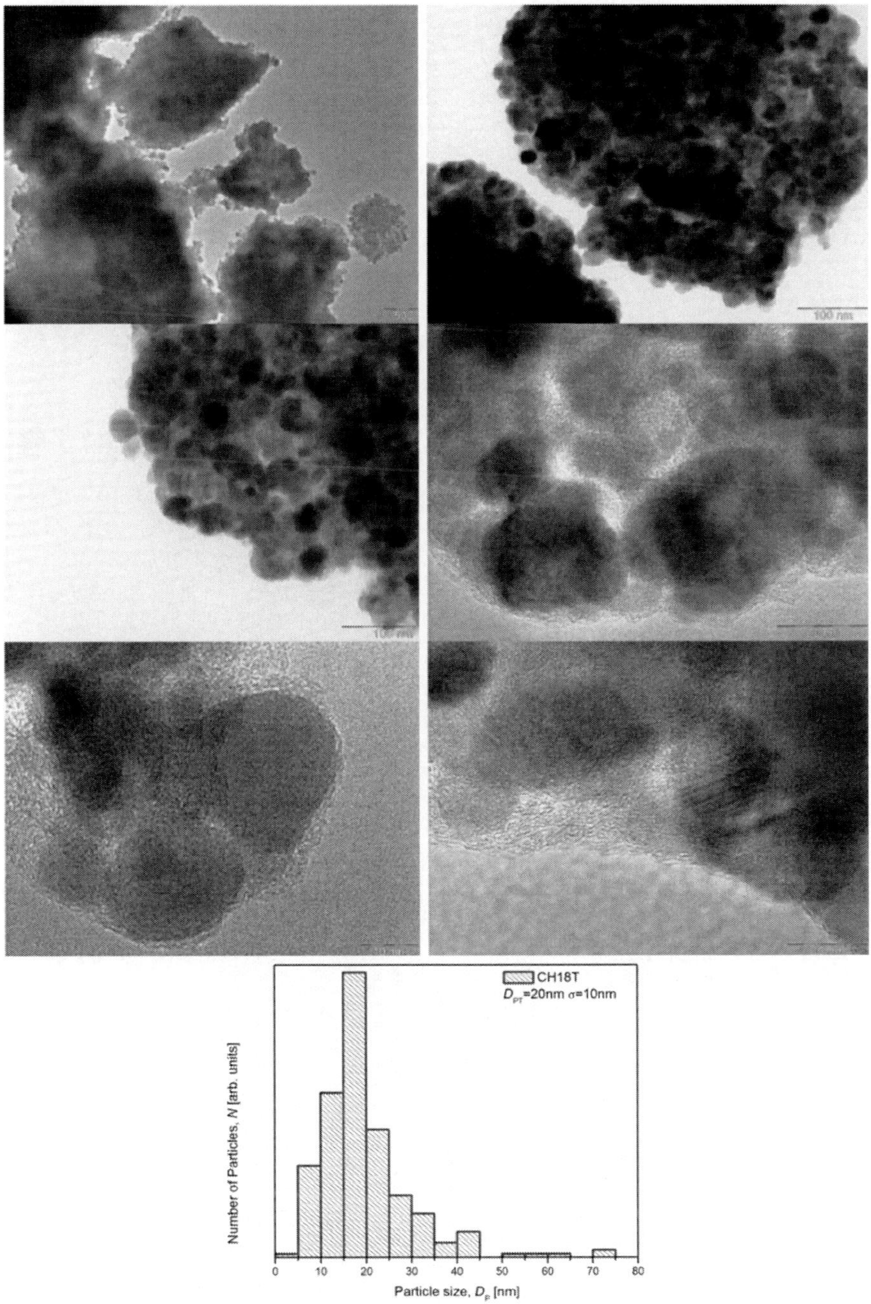

Figure 8. TEM and HRTEM images of sample CH18T. The inset shows the Lognormal particle size distribution.

In Figure 8, TEM and HRTEM images of sample CH18T also show crystalline particles coated with the amorphous phase of carbon. Only one d–spacing value was identified, associated to magnetite. The inset shows a Log–normal size distribution. This fact, together with the result from XRD indicate the existence of single–phase Fe_3O_4 NPs. The secondary phase –hematite– reduces completely to magnetite, and the thermal treatment promotes the growth of magnetite crystals from an average particle size D_{PT} = 11 nm (sample CH18) to 20 nm (sample CH18T).

Magnetic Characterization

The specific magnetization (σ) as a function of the applied magnetic field (H) for samples CHt and CHtT was measured and it is presented in Figures 9 and 10, respectively.

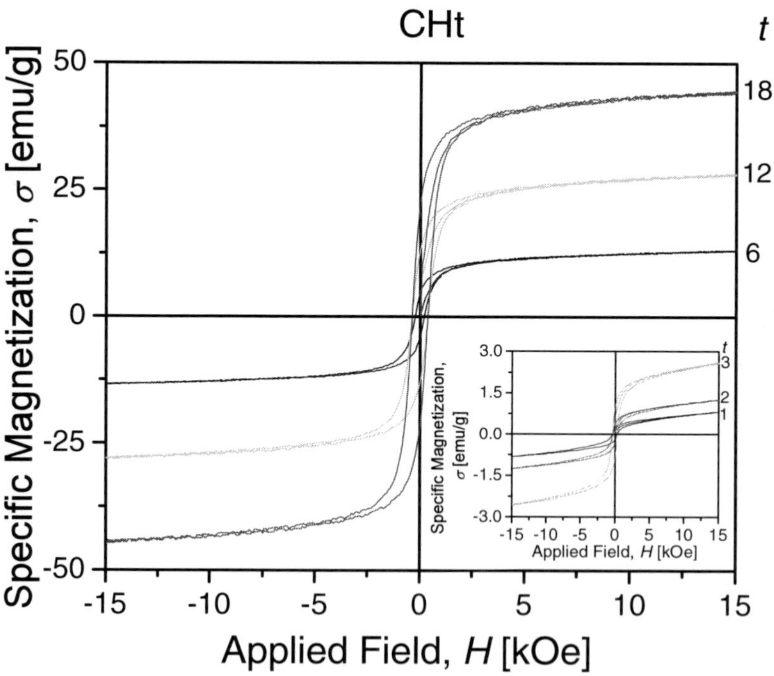

Figure 9. σ vs. H curves for samples CHt for t = 6, 12 and 18 h. The inset shows σ vs. H curves for samples CHt, for t =1, 2 and 3 h.

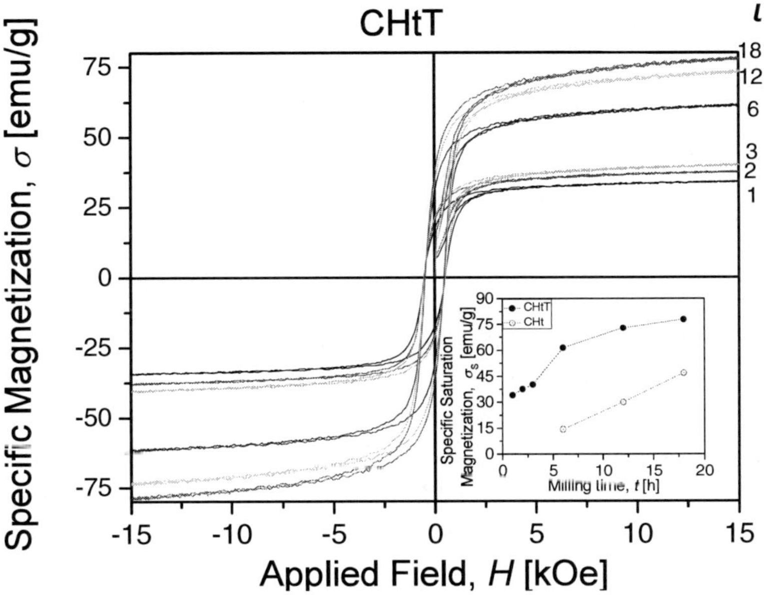

Figure 10. σ vs. H curves for samples CHtT. The inset shows σ_S values as a function of t for samples CHt and CHtT.

While the σ vs. H curves of samples CHt (for t = 1, 2 and 3 h) do not saturate and present low σ values, a significant increase in σ is observed when increasing the milling time from 1 to 3 hours (see the inset of Figure 9). This effect is due to the increase of the magnetite percentage with milling time. Samples milled from 6 to 18 hours do show a ferrimagnetic behavior (see Figure 9) as it is expected from the results obtained by XRD (Figure 1), where magnetite starts to appear after 6 h of milling and increases with milling time, t.

On the other hand, all the σ vs. H curves of samples CHtT show a typical ferrimagnetic behavior (Figure 10).

The specific saturation magnetization (σ_S) of these samples takes the values presented in the inset of Figure 10, where the σ_S values range from 34.1 emu/g after 1 h to 78.0 emu/g after 18 h (single–phase sample).

The value of specific saturation magnetization for bulk magnetite σ_S' is 92 emu/g [23], which is comparable to the specific saturation magnetization of sample CH18T (78.0 emu/g). The difference can be attributed to the fact that sample CH18T contains a fraction of carbon (non–magnetic material) apart from magnetite, so that σ_S is reduced. Besides, magnetization of a nanosized powder is usually reduced with respect to bulk magnetization values because of surface effects; in this case, the superficial chemical bonds are bound to be broken by the milling process and the annealing temperature is not high enough to restore them.

In every case, coercive fields are about 500 Oe. The powders display rather high coercivity values and a slow approach to saturation, indicating that the nanoparticles aggregate in clusters (as confirmed by SEM and TEM) with a wide dispersion of cluster dimensions.

Chantrell *et al.* [24] introduced a mathematical expression for determining the average particle size D_{PC} of a particulate system (and the corresponding standard deviation δ_C) using the magnetic parameters corresponding to σ vs. H curves, under the assumption of spherical particles with a Log–normal distribution. It is clear from the inset of Figure 8 that our systems fulfills this hypothesis; therefore Chantrell's expression can be used in this work.

Let T_r be the room temperature at the moment of the VSM measurement, χ_i the initial susceptibility, M_S the sample saturation magnetization, M'_S the bulk saturation magnetization and H_C the coercive field. Then D_{PC} and δ_C can be calculated according to [24]:

$$D_{PC} = \left[\frac{18 k_B T_r}{\pi M_S} \sqrt{\frac{\chi_i}{3 M'_S H_C}} \right]^{1/3}, \qquad (2)$$

$$\delta_C = \frac{1}{3} \cdot \ln \left(\frac{3 \chi_i H_C}{M'_S} \right)^{1/3} \qquad (3)$$

Using equations (2) and (3) with the corresponding error propagation, data from the measured σ vs H curves of Fig. 11, and the sample density from Rietveld refinements, the particle size of magnetic NPs results D_{PC} = (18±1) nm with a standard deviation δ_C=0.3, in good agreement with TEM results.

IV. Electrochemical Characterization and Biomedical Application

It is widely known that several kinds of nanomaterials exhibit enzyme–like activity in a wide range of nonphysiological conditions as extreme pH values, high temperature or in the presence of inhibitors [25]. This type of functional nanomaterials, that could be named "artificial enzymes", have attracted great attention in the last years due to several advantages as stability under harsh conditions and effective biocatalysis as replacement of natural enzymes. Magnetite NPs are not the exception, since they have demonstrated intrinsic peroxidase–like activity and have been applied as peroxidase nanomimetics to detect a wide range of biomolecules [26–32].

In order to explore this concept and characterize the electrochemical behavior of the C–coated magnetite NPs, investigate their peroxidase–like biocatalysis and consider their application in biosensing strategies, a C–coated NPs–modified composite electrode (NPs/CPE) was used in this work. For this purpose, sample CH18T was used.

For the evaluation of the electrocatalytic response towards 0.010 M hydrogen peroxide by cyclic voltammetry, three composite electrodes were assayed and compared: bare CPE (a), 5.0 % w/w commercial magnetite microparticles (MPs)/CPE (b) and 5.0% w/w NPs/CPE (c) (Figure 11 a). As shown in the figure, there is an important decrease in the overvoltage for the electro–oxidation and mainly for the electro–reduction of hydrogen peroxide and a significant increase of the associated currents for 5.0% NPs/CPE as expected, compared to the poor electrochemical response at the bare electrode (CPE) [11, 33–36] and at 5.0% MPs/CPE [36].

At the 5.0% C-coated NPs/CPE, the reduction current starts to increase at potentials more negative than 0.200V, as expected according to the excellent electrocatalytic action of magnetite NPs towards the reduction of hydrogen peroxide [25, 32, 35]. It is important to remark that NPs also catalyze the hydrogen peroxide oxidation, starting at more positive potentials than 0.300 V, although this effect is less pronounced than in the case of the reduction.

Thus, the preferential reduction of hydrogen peroxide at C–coated NPs/CPE, allows reaching low–potential values for the amperometric detection, offering a good alternative for the selective enzymatic quantification of glucose. Therefore, –0.100 V was selected as working potential for further amperometric experiments. Figure 11 b depicts cyclic voltammograms obtained for 0.010 M hydrogen peroxide at CPE containing increasing amounts of C–coated magnetite NPs: 0.0 (A), 2.5 (B), 5.0 (C), 10.0 (D) and 20.0 (E) % w/w. As expected, there is an enhancement in the reduction currents for hydrogen peroxide with the amount of nanomaterial, whereas there is an important increase in the capacitive currents that can be seen more clearly at the 20.0 %NPs/CPE. This behavior is due to the nonconductive nature of magnetite and a resistive behavior due to a poor agglutination of the composite matrix with high content of NPs, as was reported previously [11].

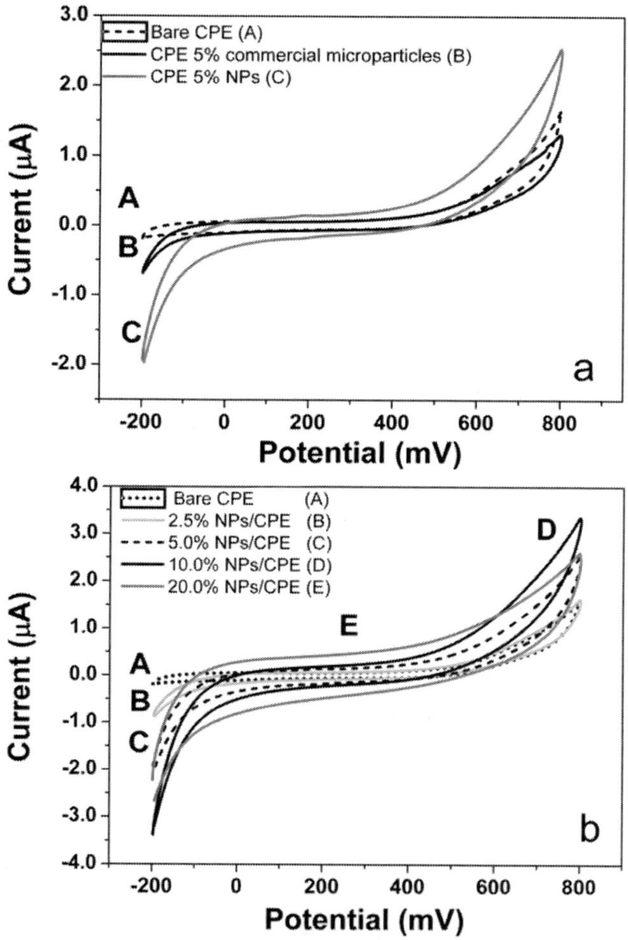

Figure 11. (a) Cyclic voltammograms for 0.010 M hydrogen peroxide at (A-dashed line) bare carbon paste electrode (CPE), at (B-solid black line) CPE modified with 5.0% w/w Fe_3O_4 microparticles (5.0% MPs/CPE) and at (C-solid grey line) CPE modified with 5.0% w/w C-coated Fe_3O_4 nanoparticles (5.0% NPs/CPE). Supporting electrolyte: 0.050 M phosphate buffer solution pH 7.40. Scan rate: 0.100 Vs^{-1}. (b) Cyclic voltammograms for 0.010 M hydrogen peroxide at CPE containing different percentages w/w of C-coated Fe_3O_4 nanoparticles: 0.0 (A-dotted line), 2.5 (B-light grey line), 5.0 (C-dashed line) 10.0 (D-solid grey line) and 20.0 (E-solid black line) % w/w NPs: Supporting electrolyte: 0.050 M phosphate buffer solution pH 7.40. Scan rate: 0.100 Vs^{-1}.

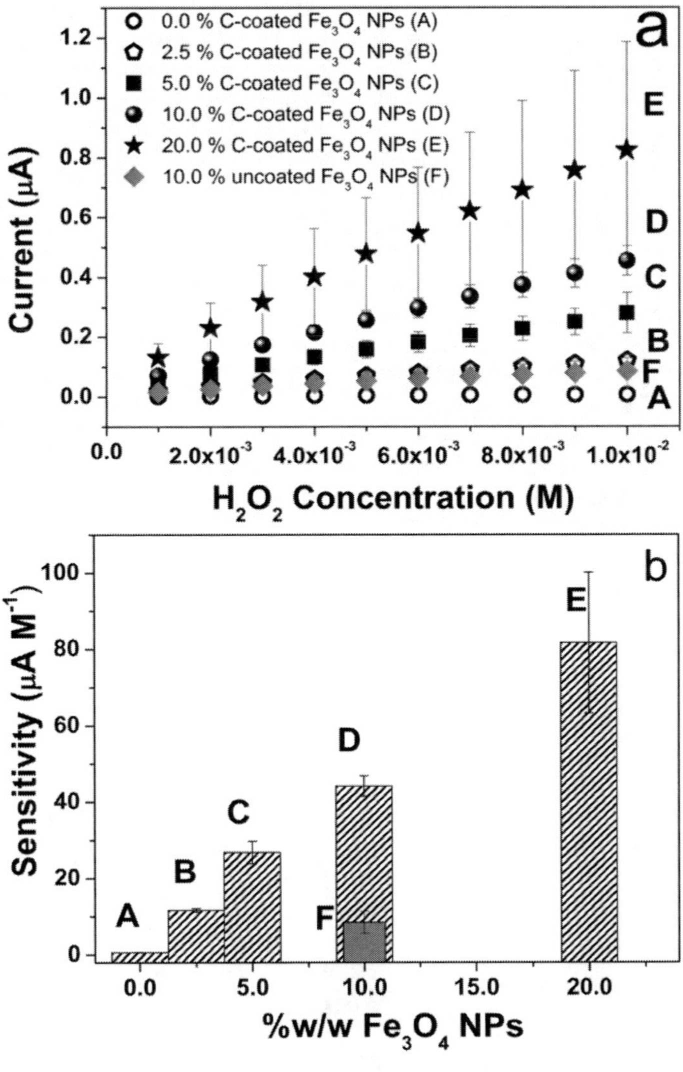

Figure 12. (a) Calibration plots obtained from amperometric experiments for successive additions of 1.0 x 10^{-3} M hydrogen peroxide performed at –0.100 V at CPE containing different percentages w/w of C-coated Fe_3O_4 nanoparticles: 0.0 (A), 2.5 (B), 5.0 (C), 10.0 (D), 20.0 (E) % w/w and uncoated Fe_3O_4 nanoparticles 10.0% w/w(F). Supporting electrolyte: 0.050 M phosphate buffer solution pH 7.40. (b) Sensitivity towards hydrogen peroxide obtained from amperometric experiments at -0.100 V (shown in a) at CPE containing different percentages w/w of C-coated Fe_3O_4 nanoparticles: 0.0 (A), 2.5 (B), 5.0 (C), 10.0 (D) and 20.0 (E) % w/w and uncoated Fe_3O_4 nanoparticles 10.0% w/w (F). Supporting electrolyte: 0.050 M phosphate buffer solution pH 7.40.

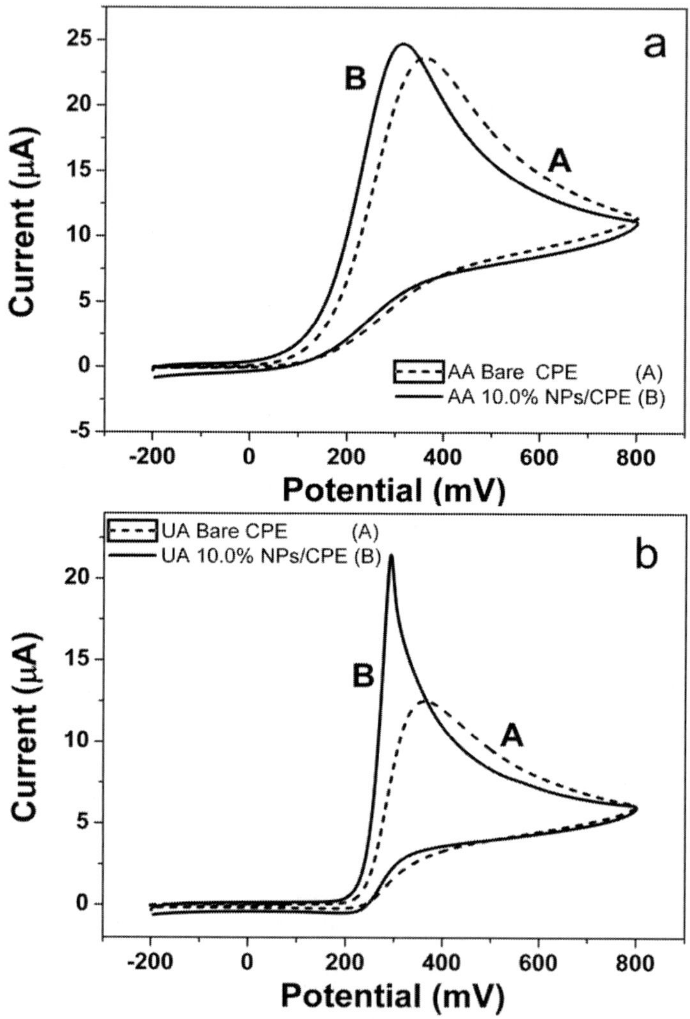

Figure 13. Cyclic voltammograms for (a) 1.0 x 10^{-3} M ascorbic acid (AA) and (b) 5.0 x 10^{-4} M uric acid (UA) at (A–dashed line) bare carbon paste electrode (CPE) and at (B–solid line) CPE modified with 10.0% w/w C–coated Fe_3O_4 nanoparticles (CPE 10.0% NPs). Supporting electrolyte: 0.050 M phosphate buffer solution pH 7.40. Scan rate: 0.100 Vs^{-1}.

The content of magnetite NPs in the composite material proved to be a very critical variable for the performance of the electrode. Figure 12 a displays calibration plots obtained from amperometric recordings performed at –0.100 V for successive additions of 1.0 x 10^{-3} M hydrogen peroxide at CPE containing different percentages of C–coated magnetite NPs: 0.0 (A), 2.5 (B),

5.0 (C), 10.0 (D) and 20.0 (E) %w/w and 10.0 (F) %w/w uncoated–magnetite NPs, respectively. Figure 12 b shows the corresponding sensitivity values towards hydrogen peroxide obtained from the calibration plots at –0.100 V (shown in Figure 12 a).

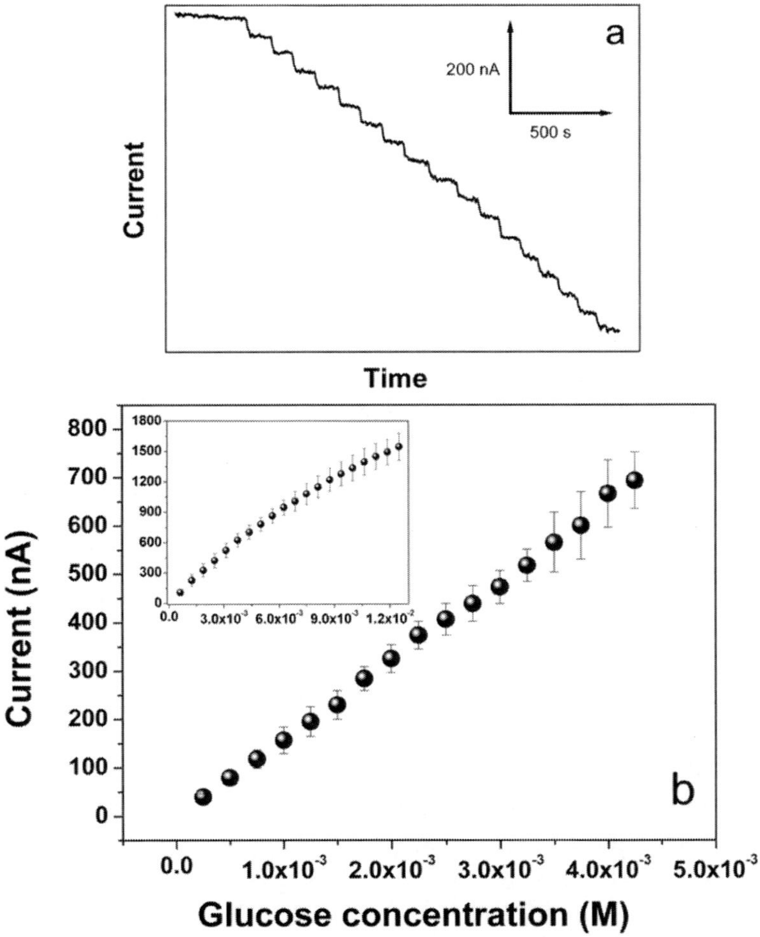

Figure 14. (a) Amperometric recordings for successive additions of 2.5 x 10^{-4} M glucose performed at -0.100 V at CPE containing 10.0 % w/w C–coated Fe_3O_4 nanoparticles and 5.0 % w/w GOx (10.0%C-coated NPs–5%GOx-CPE). Supporting electrolyte: 0.050 M phosphate buffer solution pH 7.40. (b) Calibration plot obtained from the amperometric recordings (shown in a). Inset: Calibration plot for successive additions of 6.25 x 10^{-4} M glucose in an extended range of concentrations. Working potential: -0.100 V. Supporting electrolyte: 0.050 M phosphate buffer solution pH 7.40.

The corresponding sensitivities are the following: (0.63 ± 0.07), (11.7 ± 0.5), (27 ± 3), (44 ± 3), (8 ± 2) x 10^1 μAM^{-1} for CPE containing 0.0; 2.5; 5.0; 10.0; and 20.0% w/w C–coated magnetite NPs, respectively; while for 10.0% w/w uncoated magnetite NPs the sensitivity value found was (9 ± 3) μAM^{-1} . Analyzing the sensitivity values for CPE containing increasing amount of C–coated magnetite NPs, a linear relationship is obtained (with regression coefficient R^2= 0.997). As it was previously concluded from the cyclic voltammetric experiments (Figure 11 b), there is an important influence of the content of C–coated NPs on the sensitivity towards hydrogen peroxide. However, for 20.0% C–coated NPs/CPE, in addition to the important increase of the sensitivity, a very low reproducibility of the calibration curve is observed. This fact could be attributed to the agglutination difficulty of the composite matrix with the increasing content of magnetite NPs [35]. In view of these results, a 10.0% w/w C–coated magnetite NPs was selected as the optimum composition of the electrode material for a highly sensitive response. Likewise, the sensitivities towards hydrogen peroxide at 10.0% C–coated magnetite NPs/CPE and 10.0% uncoated magnetite NPs/CPE were compared (Figure 12 a, curve D; Figure 12 a, curve F; Figure 12 b, column D and Figure 12 b, column F) under the same experimental conditions. The sensitivity obtained for C–coated magnetite NPs is higher than the one for uncoated magnetite NPs by a factor of 5, demonstrating that the carbon surrounding provides a better environment for the catalytic activity and the electron transfer of magnetite NPs.

In the development of electrochemical biosensing strategies for blood glucose determination, a very remarkable aspect to take into account is the interference of easily oxidizable compounds, usually present in biological fluids, such as ascorbic acid (AA) and uric acid (UA). For this purpose, the catalytic activity of C–coated NPs/CPE towards the oxidation of these compounds was investigated. Figure 13 shows the typical voltammetric profiles obtained at 0.100 Vs^{-1} for 1,0 x 10^{-3} M AA (a) and for 1,0 x 10^{-3} M UA (b) at CPE (A–dotted line) and 10.0 % C-coated NPs/CPE (B–solid line). For C–coated NPs/CPE, there is a negative shift of 46 mV in the oxidation peak potential for AA and a negligible increase in the peak current from 23.6 to 24.7 μA. Moreover, for UA the catalytic effect is more evident, with a negative shift of 73 mV in the oxidation peak potential and an increase in the peak current from 12.5 to 21.6 μA (1.7 times). Nevertheless, although C–coated NPs reveal a slight catalysis for the oxidation of UA, the effect is not significant compared to the catalysis exhibited for hydrogen peroxide. Furthermore, at -0.100 V —the selected potential for amperometric detection of hydrogen peroxide— there is no response to successive additions of AA and UA, even at higher concentrations than the physiological levels of these compounds (not shown), allowing a highly specific detection of hydrogen peroxide. Therefore, these results allow to conclude that C–coated magnetite NPs/CPE constitute a highly efficient system for the electrocatalytic reduction of hydrogen peroxide, making it possible to consider its application in biosensing strategies for highly sensitive and selective glucose determination.

Upon the excellent performance of C–coated magnetite NPs/CPE towards hydrogen peroxide reduction and the remarkable selectivity in amperometric experiments, the following step was the design of a glucose biosensor by modifying 10.0% NPs/CPE with glucose oxidase (GOx) as biocatalyst. GOx catalyzes the oxidation of glucose to gluconic acid in presence of oxygen which is the natural regenerator of the enzyme active site, producing hydrogen peroxide after GOx regeneration. The amount of GOx within the composite has demonstrated to be an important variable in the development of the glucose biosensor, this amount is usually comprised between 5.0 and 10.0% [11, 33]. It is important to notice that for excessive amounts of GOx, commonly higher than 10.0% w/w, the oxygen consumption becomes more important, producing a shortening of the linear range [11, 34]. Therefore 5.0% w/w GOx was selected in order to compare the proposed magnetite NPs modified composite electrode with previous results [11]. In the designed biosensing strategy, the biosensor was obtained by dispersing 5.0% w/w of GOx within CPE modified with 10.0% w/w C–coated magnetite NPs. The reduction current of the hydrogen peroxide produced during the step of GOx regeneration obtained at –0.100 V, was taken as analytical signal. Figure 14 a illustrates a current–time profile for successive additions of 2.5×10^{-4} M glucose, performed at –0.100 V at 5.0% GOx/10.0% C–coated NPs/CPE. A very fast and well–defined response is observed for each glucose addition. Similar amperometric experiments performed at 5.0% GOx/CPE at –0.100 V gave no response, as expected, due to the poor reduction of the hydrogen peroxide enzimatically generated at bare CPE [33]. Figure 14 b displays a calibration plot obtained from the current–time profile (shown in Figure 14 a), while the inset shows a calibration curve for successive additions of 6.25×10^{-4} M glucose performed at –0.100 V in an extended range of glucose concentration. A linear relationship between current and glucose concentration is obtained up to 7.5×10^{-3} M glucose (135 mg/dL), comprising the entire physiological range and incipient pathological values. For higher glucose concentrations, the current increases non-linearly, as expected for biocatalyzed reactions. The average sensitivity obtained at –0.100 V was $(1.62 \pm 0.05) \times 10^{5}$

nAM^{-1} (R^2= 0.9992) and the quantification limit (taken as the smallest measured concentration) was 1.25 x 10^{-4} M. The kinetics parameters of the bioelectrode, K_M^{app} and I_{max}, were calculated by the Eadie–Hofstee plot obtained from the data of the calibration plot shown in Figure 14 b (inset). The values obtained were the following: K_M^{app} : (0.02298 ± 0.0004) M and I_{max}: (4.39 ± 0,06) x 10^3 nA. The value found for K_M^{app} falls within the range reported for K_M of GOx in homogeneous solution [37] and in the composite ceramic carbon matrix (K_M^{app} 20 ± 2 mol/L) [38], suggesting that the environment provided by C–coated magnetite NPs/composite matrix where GOx was immobilized, is suitable for the biocatalytic activity of the enzyme.

The biosensor was further challenged by the addition of AA and UA. Figure 15 depicts the current–time profile at –0.100 V obtained after one addition of 5.0 x 10^{-4} M glucose, followed by successive additions of 2.0x10^{-5} M AA (a) and 4.0 x 10^{-5} M UA (b) up to achieve final concentrations of 2.0 x 10^{-4} M and 4.0 x 10^{-4} M, respectively. No interference signal in the presence of glucose was revealed for such concentrations, which are even higher than the maximum physiological levels found in human blood serum. This fact evidences the advantages of the preferential electrocatalytic detection of the enzymatically generated hydrogen peroxide, leading to a high sensitivity and an excellent selectivity.

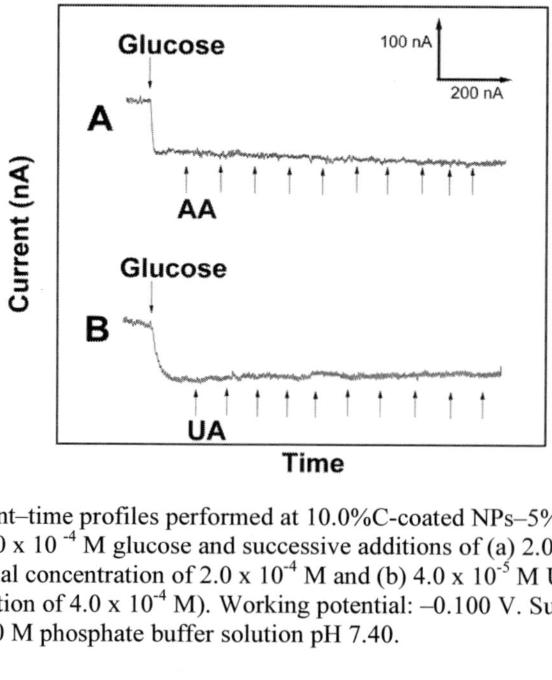

Figure 15. Current–time profiles performed at 10.0%C-coated NPs–5%GOx–CPE) for an addition of 5.0 x 10^{-4} M glucose and successive additions of (a) 2.0 x 10^{-5} M AA (up to reach a final concentration of 2.0 x 10^{-4} M and (b) 4.0 x 10^{-5} M UA (up to reach a final concentration of 4.0 x 10^{-4} M). Working potential: –0.100 V. Supporting electrolyte: 0.050 M phosphate buffer solution pH 7.40.

In order to evaluate the applicability of the bioelectrode for glucose practical determinations, 5.0% GOx/10.0% C–coated NPs/CPE was examined in the presence of human blood serum samples (Standatrol S–E–2, levels physiological N1 and pathological N2, Wiener Lab.). The glucose concentration obtained in the sample after 5 determinations were (91.6 ± 0.4) mg/dL for level N1 and (232 ± 6) mg/dL for level N2, while the glucose concentration reported by Wiener for levels N1 and N2 were 89 mg/dL and 238 mg/dL, respectively. These values gave a relative error of +2.8% and –2.6% for N1 and N2 samples, respectively, indicating the noteworthy performance of the proposed glucose biosensor for practical assays.

CONCLUSION

Stable carbon–coated magnetite NPs at room temperature were succesfully synthetized by the mechanochemical method and further thermal treatment at 500 ºC. A reduction of hematite to magnetite was induced by the mechanical process in which amorphous carbon was used in excess so as to remain covering the ferrite particles. The specific saturation magnetization of the resultant powder is very close to the bulk value. Crystal and particle sizes are the same (20 nm), indicating the obtention of single–crystal magnetite nanoparticles.

The use of C–coated magnetite nanoparticles for the development of an electrochemical glucose biosensor was proposed for the first time. C–coated magnetite NPs have demonstrated an excellent catalytic activity towards hydrogen peroxide enzimatically generated. This feature allowed quantifying glucose at potentials negative enough to avoid the interference of a large excess of easily oxidizable compounds or complex matrices such as human blood serum. The resulting bioelectrode demonstrated an excellent analytical performance, comparable and even better than previous works using magnetite nanoparticles obtained by electrochemical synthesis with surfactants, as catalytic nanomaterial [11]. The simplicity of the preparation, the high sensitivity and selectivity, the excellent reproducibility of the proposed biosensor, and the excellent correlation of the results for blood human serum with those obtained using the spectrophotometric method, convert the proposed biosensor in a very good promise for practical applications in real biological systems. Based on the catalytic activity of C–coated magnetite nanoparticles and their known possibilities of functionalization, it is expected that the surface modification of these nanoparticles with biomolecules will allow to have very efficient biosensors for a wide range of biomolecules.

This investigation is of particular significance for technological applications of C–coated NPs in biotechnology, environmental chemistry and biomedicine. It is expected that the synergic scientific contributions from different areas including chemistry, materials science, biology and clinical medicine would lead, in the near future, to further applications of modified magnetic nanomaterials in the development of smart–devices for theranostics purposes in the detection of high–impact markers of important pathological processes.

ACKNOWLEDGMENTS

This work was partially funded by Secyt-UNC, Conicet and ANPCyT.

REFERENCES

[1] Malsch, I. (2002). Biomedical applications of nanotechnology, *The Industrial Physicist*, 8(3), 15–17.

[2] Mehier–Humbert S. & Guy, R.H. (2005). Physical methods for gene transfer: improving the kinetics of gene delivery into cells, *Advanced Drug Delivery Reviews*, 57, 733–53.

[3] Podzus, P. E., Daraio M. E. & Jacobo S. E. (2009). Chitosan magnetic microspheres for technological applications: Preparation and characterization, *Physica B*, 404, 2710–2712.

[4] Aphesteguy, J. C., Jacobo, S. E., Schegoleva, N. N. & Kurlyandskaya G. V. (2010). Characterization of nanosized spinel ferrite powders synthesized by coprecipitation and selfcombustion method, *Journal of Alloys and Compounds*, 495 (2), 509–512.

[5] Sánchez, F., Mendoza Zélis, P., Pasquevich, G., Fernández van Raap, M., Stewart, S., Aphesteguy, J. & Jacobo, S. E. (2013). Structural and magnetic study of zinc- doped magnetic nanoparticles and ferrofluids for hyperthermia aplications. *Journal of Physica D: Applied Physics*. 46, 125006 (12pp).

[6] Bertoglio, P., Jacobo, S. E. & Daraio, M. E. (2010). Preparation and characterization of PVA films with magnetic nanoparticles: The effect of

particle loading on drug release behavior, *Journal Of Applied Polymer Science*, 115(3) 1859–1865.

[7] Albornoz, C., Sileo, E. E. & Jacobo, S. E. (2004). Magnetic polymers of maghemite (g-Fe2O3) and polyvinyl alcohol, *Physica B,* 354, 149–153.

[8] Albornoz, C. & Jacobo, S. E. (2006). Preparation of a biocompatible magnetic film from an aqueous ferrofluid. *Journal of Magnetism and Magnetic Materials*, 305(1), 12–15.

[9] François, N. J., Allo, S., Jacobo & S. E., Daraio, M. E. (2007). Composites of polymeric gels and magnetic nanoparticles: preparation and drug release behavior, *Journal of Applied Polymer Science*, 105, 647–655.

[10] [10] Lee, T. M. H. (2008). Over-the-counter biosensors: Past, Present, and Future *Sensors*, Vol 8(9), 5535–5559.

[11] [11] Comba, F. N., Rubianes, M. D., Cabrera, L., Gutiérrez, S., Herrasti, P. & Rivas, G. A. (2010). Highly sensitive and selective glucose biosensing at carbon paste electrodes modified with electrogenerated magnetite nanoparticles and glucose oxidase, *Electroanalysis*, 22(14), 1566–1572.

[12] Comba, F. N., Gutiérrez, F., Herrasti, P., Rubianes, M. D. & Rivas, G. A. (2012). Effect of the incorporation of proteins on the performance of carbon paste electrodes modified with electrogenerated magnetite nanoparticles towards the reduction of hydrogen peroxide, *Electroanalysis*, 24, 1541–1546.

[13] Comba, F. N., Rubianes, M. D., Herrasti, P., & Rivas, G. A. (2010), Glucose biosensing at carbon paste electrodes containing iron nanoparticles, *Sensors and Actuators B*, 149, 306–309.

[14] Lea, M. C. (1984). *Philosophical Magazine*, 34, 470–475.

[15] Suryanarayana, C. (2001). Mechanical alloying and milling, *Progress in Materials Science*, 46, 1–184.

[16] Takacs, L. (2002). Self-sustaining reaction induced by ball milling, *Progress in Materials Science*, 47, 355–414.

[17] Schaffer, G. B. & McCormick, P. G. (1990). Displacement Reactions during Mechanical Alloying, *Metallurgical Transaction A*, 21, 2789–2794.

[18] Forrester, J. S. & Schaffer, G.B. (1995). The chemical kinetics of mechanical alloying, *Metallurgical Transaction A*, 26, 725–730.

[19] Chicinas, I. (2006). Soft magnetic nanocrystalline powders produced by mechanical alloying routes. *Journal of Optoelectronics and Advanced Materials*, 8(2), 439–448.

[20] Arana, M., Jacobo, S. E., Troiani, H. & Bercoff, P. G. (2013). Synthesis and characterization of carbon-coated magnetite for functionalized ferrofluids, *IEEE Transactions on Magnetics*, 49(8), 4547–4550.

[21] Hisa M., Tsutsumi A. & Akiyama T. (2004). Reduction of iron oxides by nano-sized graphite particles observed in pre-oxidized iron carbide at temperatures around 873 K. *Materials Transaction*, 45(6), 907–910.

[22] Arana, M., Galván, V., Jacobo, S. E. & Bercoff, P. G. (2013). Cation distribution and magnetic properties of LiMnZn ferrites. *Journal of Alloys and Compounds*, 568, 5–10.

[23] Han, D. H., Wang, J. P. & Luo, H. L. (1994). Crystallite size effect on saturation magnetization of fine ferrimagnetic particles. *Journal of Magnetism and Magnetic Materials*, 136, 176–182.

[24] Chantrell, R. W., Popplewell, J. & Charles S. W. (1978). Measurements of particle-side distribution parameters in ferrofluids. *IEEE Transaction on Magnetics*, 14, 975–977.

[25] Shanhu, L., Feng, L., Ruimin, X., & Jun-Jie, Z. (2011). Structural effects of Fe_3O_4 nanocrystals on peroxidase-like activity *Chemistry A European Journal*, 17(2), 620–625.

[26] Cheng–Hao, L., Cheng–Ju, Y. & Wei-Lung, T. (2012). Fluorescence assay of catecholamines based on the inhibition of peroxidase–like activity of magnetite nanoparticles. *Analytica Chimica Acta*, 745, 143–148.

[27] Jianbo, L., Xiaona, H., Shuai, H., Tao, W., Wenqi, L., Xing, Z., Jun-Jie, Y. & Xiaochun, W. (2012). Au@Pt core/shell nanorods with peroxidase- and ascorbate oxidase–like activities for improved detection of glucose. *Sensors and Actuators B,* 166–167, 708–714.

[28] Chen, Z., Yin, J. J., Zhou, Y. T., Zhang, Y., Song, L., Song, M., Hu, S. & Gu, N. (2012). Dual enzyme–like activities of iron oxide nanoparticles and their implication for diminishing cytotoxicity. *ACS Nano*, 6(5), 4001–4012.

[29] Ai–Xian Z., Zhong-Xiao, C., Jin-Ru, W., Juan L., Huang-Hao,Y. & Guo-Nan, C. (2013). Highly–efficient peroxidase–like catalytic activity of graphene dots for biosensing. *Biosensors and Bioelectronics*, 49, 519–524.

[30] Patila M., Pavlidis, I. V., Diamanti, E. K., Katapodis P., Gournis D. & Stamatis H. (2013). Enhancement of cytochrome c catalytic behaviour by affecting the heme environment using functionalized carbon-based nanomaterials. *Process Biochemistry*, 48(7), 1010–1017.

[31] Teymourian H., Salimi A. & Khezrian S. Fe_3O_4 magnetic nanoparticles/reduced graphene oxide nanosheets as a novel electrochemical and bioeletrochemical sensing platform. (2013). *Biosensors and Bioelectronics*, 49, 1–8.

[32] Xiaoying N., Yinyin X., Yalei D., Liye Q., Shengda Q., Hongli C. & Xingguo C. (2014). Visual and quantitative determination of dopamine based on $Co_xFe_{3-x}O_4$ magnetic nanoparticles as peroxidase mimetics. *Journal of Alloys and Compounds*, 587, 74–81.

[33] Rodríguez, M. C. & Rivas, G. A. (2001). Highly selective first generation glucose biosensor based on carbon paste containing copper and glucose oxidase. *Electroanalysis*, 13 (14), 1179–1184.

[34] Luque, G. L., Rodríguez, M. C. & Rivas, G. A. (2005). Glucose biosensors based on the immobilization of copper oxide and glucose oxidase within a carbon paste matrix. *Talanta*, 66, 467–471.

[35] Erden, P. E., Zeybek, B., Pekyardımc, Ş. & Kılı, E. 2013. Amperometric carbon paste enzyme electrodes with Fe3O4 nanoparticles and 1,4-Benzoquinone for glucose determination. *Cells, Nanomedicine, and Biotechnology*, 41, 165–171.

[36] Lin M. S. & Leu H. J. 2005. A Fe_3O_4-based chemical sensor for cathodic determination of hydrogen peroxide. *Electroanalysis*, 17(22), 2068–2073.

[37] Kamin, R. A. & Wilson, G. S. (1980). Rotating ring-disk enzyme electrode for biocatalysis, cinetic studies and characterization of the immobilized enzyme layer. *Anal. Chem.*, 52, 1198-1205.

[38] Sampath, S. & Lev, O. (1996). Inert metal-modified, composite ceramic–carbon, amperometric biosensors: renewable controlled reactive layer. *Anal. Chem.* 1996, 68, 2015–2021.

In: Magnetic Nanoparticles
Editor: Nora P. Sabbas
ISBN: 978-1-63117-434-6

Chapter 3

MAGNETIC NANOPARTICLES FOR MULTIMODAL BIO-IMAGING

***Teruyuki Kondo,*[*1] *Yu Kimura,*[2] *Hisatsugu Yamada*[1] *and Akio Toshimitsu*[3]**

[1]Advanced Biomedical Engineering Research Unit,
Center for the Promotion of Interdisciplinary Education and Research,
Kyoto University, Katsura, Nishikyo-ku, Kyoto, Japan
[2]Research and Educational Unit of Leaders for Integrated Medical System,
Center for the Promotion of Interdisciplinary Education and Research,
Kyoto University, Katsura, Nishikyo-ku, Kyoto, Japan
[3]Division of Multidisciplinary Chemistry, Institute for Chemical Research,
Kyoto University, Gokanosho, Uji, Kyoto, Japan

ABSTRACT

One of the most rapidly growing and exciting applications of nanotechnology in biomedical research is the development of biocompatible magnetic nanoparticles (NPs) for use in powerful and highly complementary bio-imaging modalities. Biocompatible magnetic NPs should be effective as contrast agents (CAs) in magnetic resonance imaging (MRI), and the unique physical properties of magnetic NPs as well as their conjugation with dyes and/or radioactive elements could lead

[*] Corresponding author: teruyuki@scl.kyoto-u.ac.jp (TK).

to the development of the multimodal bio-imaging probes for use in MRI/optical imaging, PET/MRI, and MRI/optical/therapy. Multimodal bio-imaging is quite attractive for the early and precise diagnosis of disease, leading to theranostics, since it combines the advantages of, and provides complementary information from, several imaging modalities.

The metabolism of magnetic NPs strongly depends on their size, but also on individual surface characteristics, like charge, morphology and surface chemistry. Accordingly, the creation of novel magnetic NPs for multimodal bio-imaging can be realized only with the fusion of multiple disciplines, including chemistry, material science, biology, and engineering.

In this chapter, recent advances in the use of magnetic NPs, represented by iron oxide NPs, manganese oxide NPs, and gadolinium oxide NPs, for multimodal bio-imaging are highlighted and summarized.

INTRODUCTION

Significant advances are still being made in bio-imaging techniques through developments in the chemistry of imaging probes as well as the engineering of imaging modalities for the much earlier detection of various diseases, represented by cancers. However, individual imaging modalities have their own advantages and limitations (Table 1). For example, optical imaging using fluorescence is highly sensitive and shows a sub-micrometer spatial resolution. However, the absorption and scattering characteristics of tissue components limit the penetration depth into tissue to typically less than 1 cm. With the use of luminescence, the penetration depth improves to less than 5 cm, but the spatial resolution is not satisfactory. Positron emission tomography (PET) is a highly sensitive radionuclide-based imaging technique, but it has a limited spatial resolution because of the size limitation of the scintillator. Magnetic resonance imaging (MRI) and computed tomography (CT) represent an almost irresistible marriage of convenience and high spatial resolution, but are comparatively insensitive methods of imaging. Ultrasound (US) imaging is also widely used for imaging of the heart, liver, many abdominal organs, and vascular system because of its noninvasiveness and low cost. However, the objectivity of an US image with contrast agent (usually with microbubbles or nanoemulsions) is slight and its application is limited to soft tissues.

Photoacoustic tomography (PAT), also known as laser optoacoustic imaging, has recently been developed as an emerging noninvasive, nonionizing, and deeply penetrating imaging modality [1, 2] which combines the sensitivity of optical methods with the resolution of diffraction-limited ultrasound. However, the penetration depth is limited to less than 3 cm, which is still much deeper than that with optical imaging.

Thus, multimodal bio-imaging is quite attractive for the early and precise diagnosis of disease, leading to theranostics, since it integrates the advantages of, and provides complementary information from, several imaging modalities.

In these imaging modalities, nanoparticles (NPs) have been used as probes because their size offers many advantages, such as unique magnetic and acoustic properties (called the "size effect"), avoidance from renal excretion, and penetration into various biological barriers [3, 4]. The term "size effect" refers to the preparation of particles with a single-domain configuration to exhibit superparamagnetism [5] or with a large heat capacity to generate ultrasound.

In this chapter, recent advances in the use of magnetic NPs, represented by iron oxide NPs, manganese oxide NPs, and gadolinium oxide NPs, for multimodal bio-imaging are highlighted and summarized.

Table 1. Sensitivity and resolution of existing in vivo imaging modalities

Imaging modality	Imaging nanoparticulate agent	Resolution	Sensitivity	Advantages	Disadvantages
Optical					
Luminescence	Silver or lanthanide-doped NPs	+	+++	Sensitivity	Limited depth to less than 5 cm, low resolution
Fluorescence	QDs, dye-doped NPs, SWNTs and other carbon-based nanomaterials	+	+++	Sensitivity	Limited depth to less than 1cm, high S/N ratio for some probes
Radioactive					
PET	NPs containing positron-emitting nuclides (^{18}F, ^{11}C, ^{64}Cu, ^{124}I)	+	+++	Sensitivity	Access to cyclotron required, low resolution
SPECT	NPs containing gamma-emitting nuclides (^{99m}Tc, ^{111}In)	+	+++	Sensitivity	Low resolution
MRI	Iron oxide NPs, Gd(III)-doped NPs, NP-based CEST and hyperpolarized probes (^{129}Xe)	+++	+	High resolution	Low sensitivity
CT	Iodinated NPs, gold NPs, iron oxide-doped nanomaterials	+++	+	High resolution	Low sensitivity
Ultrasound	Microbubbles, nanoemulsions, silica NPs, polystylene NPs	+++	+	Low cost	Limited used for soft tissue
PAT	Gold nanoshells, nanocages, nanorods, NPs, SWNTs, dye-doped NPs	++	++	Sensitivity	Limited depth to less than 3 cm

SPECT, Single photon emission computed tomography; CT, computer tomography; PAT, photoacoutsic tomography

2. MAGNETIC NPS FOR MULTIMODAL BIO-IMAGING

2.1. Overview of the Synthetic Methods

Methods for the synthesis of magnetic NPs can be classified into several strategies, as shown in Table 2 [5, 6]. These include co-precipitation, thermal decomposition of organometallic complexes, a polyol process under reducing conditions for magnetic oxide NPs, micelle preparation, and hydrothermal/solvothermal synthesis, which can be summarized briefly as follows.

Table 2. Synthetic methods of magnetic NPs (Adapted from [5])

Method	Synthesis	Reaction temp. (°C)	Reaction period	Solvent	Size distribution	Shape control	Yield
co-precipitation	simple and ambient conditions	20-90	minutes	water	relatively narrow	not good	high/ scalable
thermal decomposition	complicated, inert atmosphere	100-320	hours-days	organic	very narrow	very good	high/ scalable
polyol process	complicated, inert or ambient atmosphere	150-200	hours	polyol	very narrow	good	high
micelle preparation	complicated, ambient conditions	20-50	hours	organic	relatively narrow	good	low
hydrothermal/ solvothermal	simple, high pressure	220	hours-days	water-ethanol	very narrow	very good	medium

Co-precipitation is one of the most common methods for synthesizing magnetite (Fe_3O_4) NPs, maghemite (γ-Fe_2O_3) NPs, and other-metal mixed ferrite (MFe_3O_4) NPs by the treatment of Fe(II) and/or Fe(III) precursors with other metal salts (M(II)) under aqueous alkaline conditions. The sizes, morphologies, and compositions of NPs vary with the preparation conditions (iron source, reaction temperature, ratio of Fe^{2+}/Fe^{3+}, pH and ionic strength) with high reproducibility [7]. While this procedure is common and convenient for the preparation of various iron oxide NPs, the relatively poor control of NP size is a major limitation.

Thermal decomposition is also a useful method for preparing iron oxide or other-metal oxide NPs. $M(CO)_n$ is suitable for the preparation of metal NPs, while the thermal decomposition of $M(acac)_n$ (acac = acetylacetonate) or M-oleate with a surfactant like fatty acids, alkylamines, or alkanethiols is suitable for the preparation of metal oxide NPs. This method provides highly crystalline magnetic NPs with proper reproducibility and a narrow size distribution [6].

The polyol process [8] is a typical method for preparing magnetic metal oxide NPs in high-boiling-point alcohols. This method can be used to synthesize lanthanide oxide NPs with a homogeneous dispersion in the alcohol and a distribution within a very narrow range of sub-5 nm sizes.

Micelle preparation refers to *in situ* co-precipitation in nanomicelles composed of a metal salt solution in a non-polar hydrocarbon (e.g., cyclohexane or xylene) with surfactants. For example, co-precipitation of ethanolic $FeCl_2/Fe(NO_3)_3$ solution in xylene with sodium dodecylbenzenesulfonate (NaDBS) under reflux is used for the large-scale synthesis of magnetic magnetite (Fe_3O_4) NPs [9]. The size of NPs can be controlled by independently changing the metal concentration and the surfactant concentration.

Hydrothermal/solvothermal synthesis is a method for preparing various nanomaterials at high temperature above the boiling point of the solvent under a high pressure in an autoclave. This method increases the reagent solubility in the solvent to give high-crystalline inorganic nanomaterials.

In addition, the stabilization of NPs in water for *in vivo* bio-imaging is usually achieved by surface modification using amphiphilic surfactants, hydrophilic ligands through chemical conjugation, hydrophobic interaction, and/or poly-ionic interaction [5].

Table 3. Surface functionalization of synthesized NPs

For	Method	Reagent, reactions
in situ covered NPs	ligand exchange	various carboydrates, proteins, synthetic polymers such as PEG, polyvinyl alcohol, polyvinylpyrrolidone, polyethyleneimine, and their block copolymers
	conjugation physically with hydrophobic interaction	alkylphosphonate surfactants, ethoxylated fatty aclcohols, phospholipids, 11-aminoundecanoate, cetyl trimethylammonium bromide (CTAB), amphiphilic polymers from PEG, polysaccharides, and pluronic PF127 copolymers
uncovered NPs	surface grafting	carboxylate, phosphonate, silane, catecol, and hydoxamic acid groups
		inorganic coating with silica, gold, other metals, and metal oxides

In general, the surface of metal oxide NPs is covered with hydroxyl groups at the terminal end. On the other hand, various surfactants cover NPs, and their surface is hydrophobic after they are prepared in non-polar solvents by thermal decomposition, micelle preparation or hydrothermal/solvothermal synthesis. There are various approaches for the surface modification of NPs to stabilize them in water as well as to introduce bioactive reagents (such as antibodies, drugs, and/or other targeting molecules) and other materials for another imaging modality from magnetic-core NPs (magnetic-core NPs are mainly for MRI). Accordingly, amphiphilic molecules, polymer ligands, coupling agents, silica and other bi-functional molecules are generally used for the surface modification of NPs, as shown in Table 3 [10, 11].

2.2. Iron Oxide NPs

Iron oxides are current some of the most common contrast agents (CAs) and are used as bio-imaging probes in nanoparticulate form for clinical imaging of the liver on T2- or T2*-weighted MRI [12] (e.g., Ferumoxide (Ferridex®) or Ferrixan (Resovist®, Cliavist®)). The multimodal use of iron oxide NPs has been reviewed by Louie [13]. Multimodal bio-imaging probes that use iron oxide NPs are summarized in Table 4.

Typically in Table 4, Yang succeeded MRI/optical dual-modality imaging of pancreatic cancer by using multifunctional iron oxide NPs conjugated with fragment of urokinase plasminogen activator receptor (uPAR; targeting molecule) and Cy5.5 (fluorophore) [32]. From the results of time-dependent optical imaging, accumulation of the NPs in tumor-inoculated region was clearly indicated and no accumulation in the liver, spleen, and many other normal organs was observed. Also they challenged the evaluation of sensitivity about tumor detection at size, the NPs showed that was able to detect tumors as small as 0.5 to 1 mm^3 by *ex vivo* optical imaging. It would have a potential for preoperative detection and localization of pancreatic cancer lesions by MRI, following intraoperative delineation of tumor margins and peritoneal metastases by optical imaging (Figure 1).

Most recently, de Rosales developed radiolabeled (^{99m}Tc) superparamagnetic iron oxide NPs (SPIONs) anchored with polyethylene glycol (PEG) as a "stealth" agent in the body for T1-weighted MRI and SPECT imaging [66], in which PEG chain was anchored to SPIONs through bisphosphonate to increase their circulation in the blood over 50 min ($t_{1/2}$ = 2.97 h) and succeeded in the *in vivo* imaging of mouse heart by both MRI and SPECT-CT (Figure 2).

Table 4. Modification of iron oxide NPs for multimodal bio-imaging (kindly adapted with permission from [13]. Copyright 2014 American Chemical Society.)

Probes	Modalities	Size (nm)	Imaging properties	Refs
			Iron Oxide Nanoparticle Conjugates	
rhodamine B	MRI/optical	13.3	attract to magnet; $\lambda_{em} = 532$ ($\lambda_{ex} = 350$), $\lambda_{em} = 572$ ($\lambda_{ex} = 850$), QY 0.13	[14]
		108	superparamag, no hysteresis (300 K), measured ΔB of cells; $\lambda_{em} = 577$ ($\lambda_{ex} = 555$)	[15]
AlexaFluor	MRI/optical	80	mag separation; Microscopy, FACS, 150 AlexaFluor/iron particle	[16]
		64.8	MRI cells; microscopy, FACS	[17]
dansyl	MRI/optical	39	$r_1 = 41.2$ mM^{-1}s^{-1}, $r_2 = 110.6$ mM^{-1}s^{-1} (3 T); $\lambda_{em} = 355$ ($\lambda_{ex} = 530$)	[18]
indolequinone	magnetic/optical	9.5	magnetic ND, used as quencher; $\lambda_{em} = 515$ ($\lambda_{ex} = 495$)	[19]
diarylethene	MRI/switchable optical	7	superparamag (300 K); $\lambda_{em} = 460/490$ ($\lambda_{ex} = 312, 410$)	[20]
oligothiophene	magnetic/optical	17.9-26.9	attract to magnet; $\lambda_{em} \approx 510, 620$ ($\lambda_{ex} = 365$)	[21]
Cy5.5, no target	MRI/optical	31	$r_1 = 23$ mM^{-1}s^{-1}, $r_2 = 59$ mM^{-1}s^{-1} (0.47 T); imaging 633ex, 670LP, 695BP, cell and rat glioma imaging	[22-24]
Cy5.5, target VCAM-1	MRI/optical	31	VCAM-2 targeting, microscopy, FACS, MRI	[25]
Cy5.5 and annexin V	MRI/optical	31	annexin V conjugates, cell FACS, cell pellet MRI (4.7 T), 9.4 T imaging, coronary occlusion mice, microscopy	[26, 27]
Cy5.5, target uMUC-1	MRI/optical	35.8	EPPT peptide conjugate, $r_1 = 26.43$ mM^{-1}s^{-1}, $r_2 = 53.44$ mM^{-1}s^{-1} (0.47 T); imaging 633ex, 700LP	[28]
Cy5.5. and RGD	MRI/optical	36	$r_2 = 118$ mM^{-1}s^{-1}; RGD conjugates, 8.4 Cy5.5/particle	[29]
Cy5.5 and siRNA	MRI/optical/therapy	n.d.	cell imaging	[30]
enzyme-activated Cy5.5	MRI/optical	62-68	$r_1 = 27.8\text{-}29.9$ mM^{-1}s^{-1}, $r_2 = 91.2\text{-}92.5$ mM^{-1}s^{-1}, 1.2-1.8 Cy5.5/particle	[31]
Cy5.5, target uPAR	MRI/optical	10-15	imaging pancreatic cancer mouse model, 3 T MRI, Kodak *in vivo* optical imaging ($\lambda_{ex} = 625$, $\lambda_{em} = 700$)	[32]
Cy5.5 and chlorotoxin	MRI/optical	10	MRI cells (4.7 T), microscopy	[33]
^{64}Cu-DOTA	MRI/PET	45	superparamag, no coercivity; MRI 3 T, PET	[34]
		32	$r_1 = 14.46$ mM^{-1}s^{-1}, $r_2 = 72.55$ mM^{-1}s^{-1} (1.5 T, 37 °C); up to 22% radiolabel yield	[35]

Table 4. (Continued)

Probes	Modalities	Size (nm)	Imaging properties	Refs
^{64}Cu-DTPA	MRI/PET	20	r_1= 29 $mM^{-1}s^{-1}$, r_2 = 60 $mM^{-1}s^{-1}$ (0.47 T, 39 °C); MRI 7 T, PET-CT	[36]
^{111}I-DOTA	MRI/therapy	20, 30, 100	cell studies, whole body autorad, pharmacokinetics	[37]
^{124}I-CLIO	MRI/PET	32	MRI solution and animal 1.5 T, PET R4 rodent (Concorde Microsys)	[38]
^{18}F-PEG$_3$ and Vivotag	PET/CT	30	PET-CT phantom and mouse	[39]
			Core/Shell Iron Oxide Nanoparticles	
FITC, silica shell	MRI/optical	150	imaging only, cells, eye vein inject mice, 4.7 T MRI, microscopy	[40]
		50	imaging only, cells, labeled hMSC to mouse, 1.5 T MRI, microscopy	[41]
rhodamine, silica shell	magnetic/optical	13.7	paramagnetic (ZFC), λ_{em} = 552 (λ_{ex} = 500)	[42]
		100-150	superparamag, no remanence rt, λ_{em} = 560 (λ_{ex} = 520)	[43]
		1.3-34	superparamag, no remanence rt, microscopy (rhodamine)	[44]
		30-80	hysteresis (temp not given), microscopy cells (RITC, FITC)	[45]
		n.d.	imaging only, ex vivo 14.1 T MRI, microscopy	[46]
terbium complexes, silica shell	MRI/optical	52	superparamag, no remanence, λ_{em} = 544 (λ_{ex} = 265)	[47]
Ru(bpy), silica shell	MRI/optical	20	superparamag, no remanence rt, r_2 = 30.4 $mM^{-1}s^{-1}$ (FePt core), r_2 = 26.1mM^{-1} s^{-1}, (Fe_2O_3 core), $\lambda_{em} \approx$ 610, abs = 458, 522, 782	[48]
pyrene, silica shell	magnetic/optical	330	superparamag, no hysteresis rt; λ_{em} = 375, 385, 394 (λ_{ex} = 388)	[49]
quantum dot, silica shell	magnetic/optical	170-200	superparamag, no remanence rt; λ_{em} = 537 (QD), Au abs = 518	[50]
DiI, DiR, PAA shell	MRI/optical	88-100	r_2 = 202-208 $mM^{-1}s^{-1}$ (0.47 T, 37 °C); λ_{em} = 595, abs = 555	[51]
pyrene, MPTMS shell	magnetic/optical	547	superparamag, no remanence; λ_{em} = 375, 385, 394 (λ_{ex} = 338)	[49]
quantum dot, PAA shell	MRI/optical/therapy	250	no remanence rt, attract to magnet; λ_{em} = 593 (pH 7.4) (582, pH 2)	[52]
CdSe shell	MRI/optical	>4	superparamag (not shown), blue, green, and orange	[53]
		12-15	paramag, no remanence 298 K; λ_{em} = 550, QY 18-20%	[54]

Probes	Modalities	Size (nm)	Imaging properties	Refs
CdSe/ZnS shell	magnetic separation/optical	30	attract to magnet; $\lambda_{em} = 550$	[55]
		25	moderate magnetization (2.28 emu/g); $\lambda_{em} = 565$, abs = 530	[56]
Eu:GdO_3 shell	magnetic/optical	400	linear (magnetization vs mag field); $\lambda_{em} = 615$ ($\lambda_{ex} = 260$)	[57]
LaF_3:CeTb shell	magnetic/optical	30	attract to magnet; $\lambda_{em} = 319/490/543/584/619$ ($\lambda_{ex} = 270$), abs = 543	[58]
Y:Er:NaF_4 shell	magnetic/optical	68	superparamag, zero coercivity; $\lambda_{em} = 539/658$ ($\lambda_{ex} = 980$)	[59]
Y_2O_3:Eu shell	magnetic/optical	100	superparamag, zero coercivity rt, 4.3 emu/g; $\lambda_{em} = 610$ ($\lambda_{ex} = 260$)	[60]
gold shell	MRI/optical/therapy	18/250	$r_1 = 6.87$ $mM^{-1}s^{-1}$, $r_2 = 28.15$ $mM^{-1}s^{-1}$ (7 T, 25 °C)	[61]
			imaging only 4.7 T; abs ≈ 540, 700 nm for therapy	[62]
graphite shell, on FeCo core	MRI/optical	30	superparamag rt, $r_1 = 70$ $mM^{-1}s^{-1}$, $r_2 = 644$ $mM^{-1}s^{-1}$ (7 nm core), $r_1 = 31$ $mM^{-1}s^{-1}$, $r_2 = 185$ $mM^{-1}s^{-1}$ (1.5 T); abs UV, 808 nm for therapy	[63]
Texas Red-DPPE, Bodipy FL C5-HPC, rhoda-DPPE, lipid shell	MRI/optical	21	iron content only; microscopy	[64]
Doped Iron Oxide Nanoparticles				
terbium doped	MRI/optical	13	superparamagnetic, zero coercivity rt; $\lambda_{em} = 490/545/587/612$ ($\lambda_{ex} = 235$)	[65]

n.d. not determined

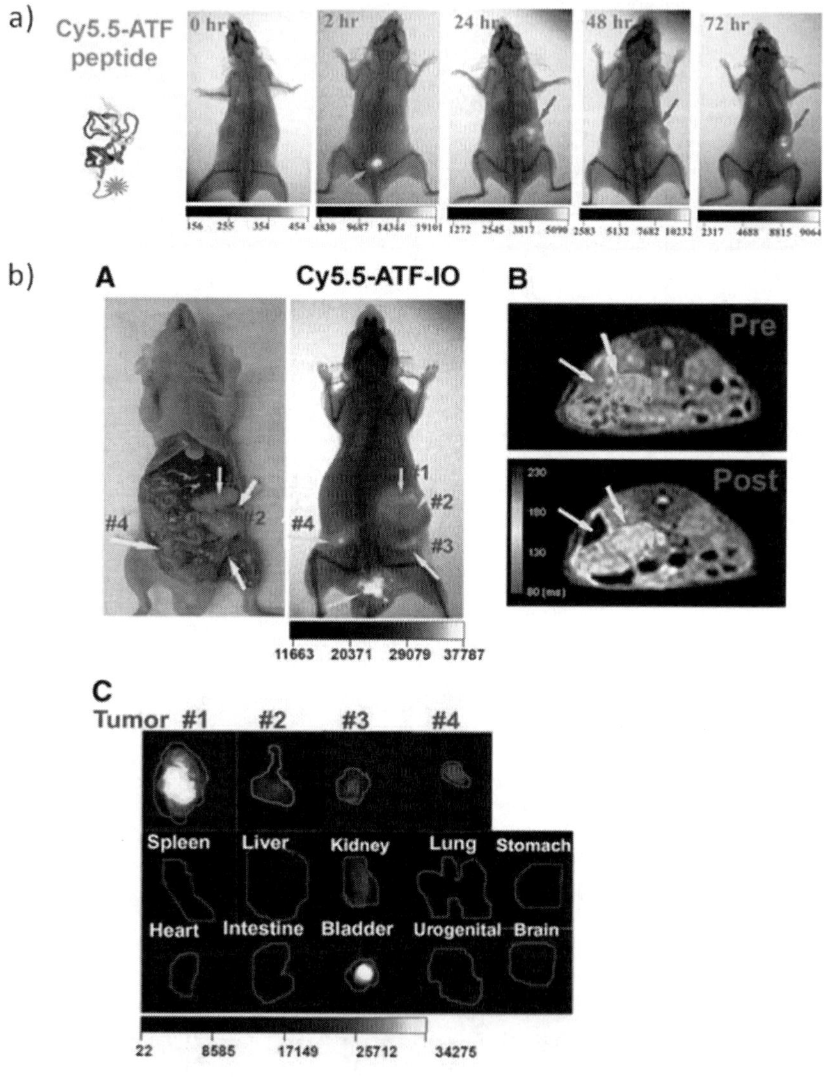

Figure 1. (Reprinted from [32]. Copyright 2014, with permission from Elsevier.) a) Examination of target spscificity of Cy5.5-uPAR fragment-iron oxide NPs using *in vivo* optical imaging in mice bearing orthotopic pancreatic cancers. Blue arrow indicates bladder and pink arrow indicates pancreatic tumor. b) Dual optical and MR imaging of pancreatic cancer using Cy5.5-uPAR fragment-iron oxide NPs. A: 48 hr after intravenous injection of NPs. Yellow arrows indicate primary (tumors #1 and #2) or metastatic (#3; 1 mm^3 and #4; 0.5 mm^3) pancreatic cancer lesions. B: T2 map obtained from the same mouse. A marked T2 reduction in the pancreatic cancer lesion (orange-red: lower T2). Blue arrow indicates pancreatic tumor and white arrow indicated spleen. C: *Ex vivo* optical imaging of tumor and normal tissues collected.

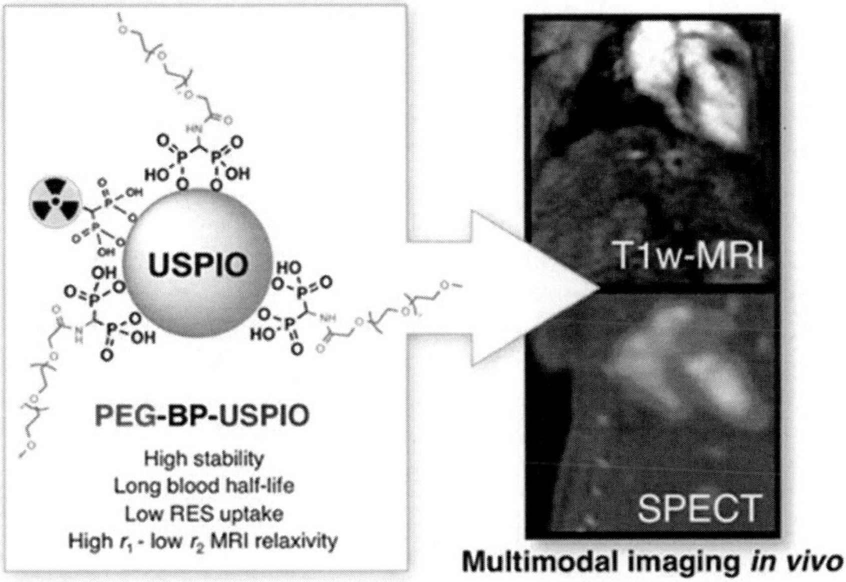

Figure 2. (Reprinted with permission from [66]. Copyright 2014 American Chemical Society.) Concept and results of bisphosphonate-anchored PEGylation and radiolabeling of iron oxide NPs. The NPs enhance the signal of blood stream 6-fold in T1-weighted MRI and also were enabled to detect by SPECT.

2.3. Manganese Oxide NPs

Although the aforementioned CAs for T2- or T2*-weighted MRI using iron oxide NPs have received considerable attention for multimodal bio-imaging, artifacts from the magnetic susceptibility of iron oxide NPs and the negative contrast effect itself often limit their much broader application in MR bio-imaging. Therefore, the study of T1-weighted MRI CAs for bright and positive contrasts has been continued.

Manganese was initially investigated because it has 5 unpaired electrons and is suitable for T1-weighted MRI [67]. Although two types of manganese(II)-based CAs were previously approved for clinical use (Mn-DPDP (Teslascan®) and $MnCl_2$ (LumenHance®)), they have recently been replaced by gadolinium(III)-based CAs. Recently, a correlation between the lavish administration of gadolinium(III)-based CAs and the pathogenesis of nephrogenic systemic fibrosis (NSF) has been reported in patients with severe renal disease, and manganese(II)-based CAs are being considerate as an alternative [68]. Typically, manganese oxide (MnO [69, 70] or Mn_3O_4 [71]) NPs and lanthanum-doped manganite ($La_{1-x}Ag_xSr_xMnO_3$) [72, 73] are studied as an candidate of T1-weighted MRI CAs.

Under these circumstances, only a few studies have considerate their multimodal used. Ken-Tye [74] and Lin [75] reported the synthesis of manganese-doped quantum dots (QDs) and the imaging of pancreatic cancer by MRI and fluorescence imaging. Choi synthesized heterostructured NPs for MRI and optical imaging, which were composed of silver/gold-MnO prepared by thermal decomposition of mixture of manganese-oleate and silver/gold-oleylamine, and following surface coating with surfactants [76]. Ha developed biotin-phospholipid-conjugated hollow manganese oxide NPs for MRI and radioactive imaging through streptoavidin-biotin conjugation chemistry, and also reported their further conjugation with anti-epidermal growth factor receptor (EGFR) antibody [77] (Figure 3).

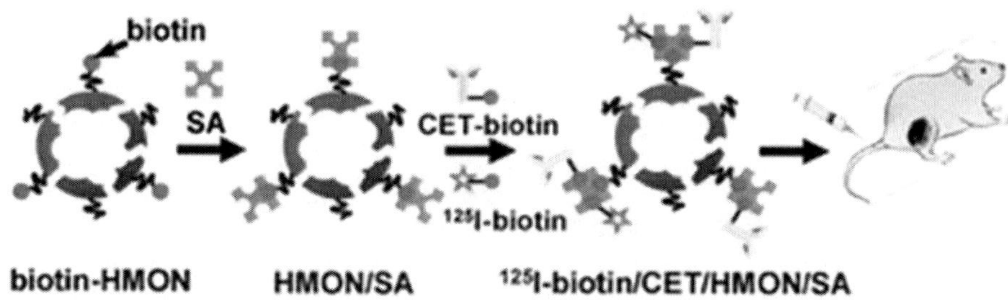

Figure 3. (Reproduced from [77] with permission of The Royal Society of Chemisty.) Conjugation techniques of radioactive spiecies (^{125}I) and targeting molecule (anti-EGFR antibody: Cetuximab; CET) toward hollow manganese oxide NPs through streptoavidin (SA)-biotin interaction for MRI and radioactive imaging.

2.4. Gadolinium Oxide NPs

Gadolinium is popular as a paramagnetic agent that bears 7 unpaired electrons for T1-weighted MRI CAs. While gadolinium(III) chelates are clinically available, their inadequate administration sometimes causes nephrogenic systemic fibrosis (NSF) and a decrease in the required dosage is a current field of study. On the other hand, gadolinium NPs may be suitable for use as MRI CAs due to their paramagnetism and stability, and Gd-based inorganic nanoparticles (NPs) such as Gd_2O_3 [78, 79] and GdF_3 [80] have recently been investigated as potential positive CAs in T1-weighed MRI. More recently, gold-speckled Gd_2O_3 NPs [81], gadolinium-conjugated Au nanorods [82], rare earth-doped Gd_2O_3 nanoplates/nanorods [83-85], and Gd-single wall carbon nanotubes (Gd-SWNTs) [86] have been used for both MRI and optical imaging. Although many examples of multimodal bio-imaging probes for MR/optical, PET/MR, and MR/optical/therapy have been reported [13, 81, 87-89], there have been no reports on the multimodal use of Gd_2O_3 NPs that are not conjugated with gold, rare earth metals, and/or dyes.

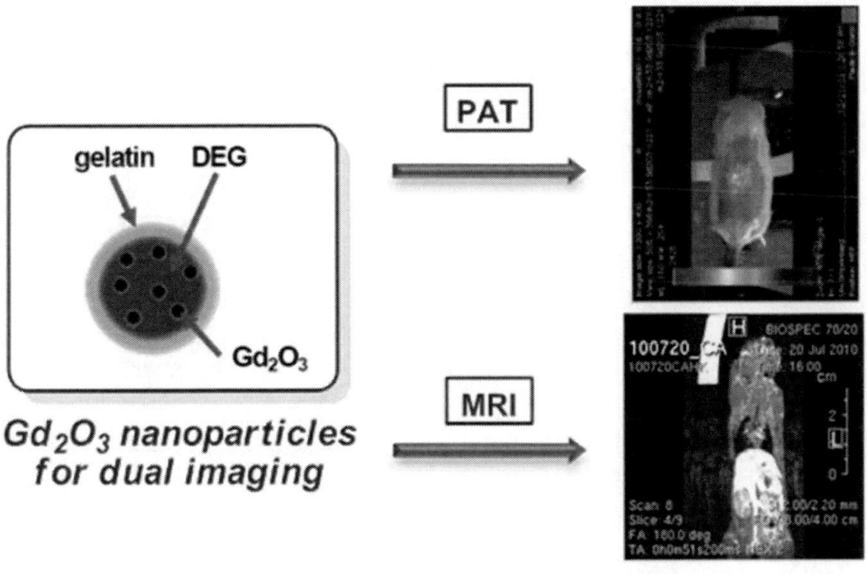

Figure 4. (Reprinted from [90].) Schematic image of gelatin-coated Gd_2O_3 NPs and their PAT and MRI image after subcutaneous and intravenous injection.

Most recently, the authors reported the synthesis of novel Gd_2O_3 NPs for MRI and PAT imaging [90] (Figure 4). A simple method for the synthesis and isolation of new size-controlled and biocompatible Gd_2O_3 NPs as a bimodal bio-imaging probe for MRI and PAT was realized by a modified polyol process starting from $Gd(NO_3)_3{\cdot}6H_2O$. The size (diameter) of the NPs could be freely controlled in the range of 20 to 200 nm by timing the addition of a biocompatible gelatin molecule to form protective colloids. NPs with a uniform diameter of around 100 nm were prepared and used in this study, since they are suitable for cellular uptake, biocompatibility and eventual elimination from the body [91], as well as passive targeting to tumor tissues through an "enhanced permeability and retention (EPR)" effect [92]. In addition, phosphates [93], lipids [13, 81], organic acids [94], sugars [95], and silica [96] were used to stabilize and solubilize Gd_2O_3 NPs in water; however, neither complete control over the diameter of NPs nor their use as bimodal bio-imaging probes for MRI and PAT imaging has been reported.

PAT in this study has a z –axis resolution of 0.3 mm, which is much less than the slice thickness in MR imaging (2 mm). Thus, it is clear that the combination of these two imaging modalities would improve the detection of lesions in superficial regions *in vivo*. Relaxivity measurement studies with these new gelatin-coated Gd_2O_3 NPs by MRI showed that the relaxivity r_1 was twice that of clinically available Gd-DTPA (Magnevist®), and that they freely circulate in the blood vessels without undesirable accumulation in the lungs. No toxicity was observed in either *ex vivo* or *in vivo* examinations.

2.5. Other NPs

Other approaches to preparing magnetic NPs as multimodal bio-imaging probes include mainly the use of liposomes which contain SPIONs and fluorophores [97]. Microbubbles that incorporate magnetic NPs as a bimodal bio-imaging probe for MRI and optical imaging have been synthesized [98]. Ran reported the synthesis of superparamagnetic perfluorooctyl bromide NPs (MPFOBNP) for US/MRI/CT imaging [99]. At present, the only examples are still limited to the use of liposomes or microbubbles as multimodal bio-imaging probes composed of low-molecular-weight paramagnetic chelates or fluorophores.

Also as an example of heterostructures for multimodal bio-imaging, "nanowontons" are reported by Bouchard (Figure 5) [100]. They prepared gold-cobalt-gold sandwich structure, called "nanowonton", by deposition of Au and Co on silicon pillar and following the etching of Si. In this structure, Au would be worked not only as a protection layer of cobalt from oxidation, but also as an optical absorbent for PAT with tunable property by changing the aspect ratio. *In vivo* MRI and *ex vivo* PAT (phantom imaging) indicated the usefulness of the probe for the detection of trace amounts (pM sensitivity) in biological tissues, in which MRI enabled the volume detection and PAT enabled the edge detection.

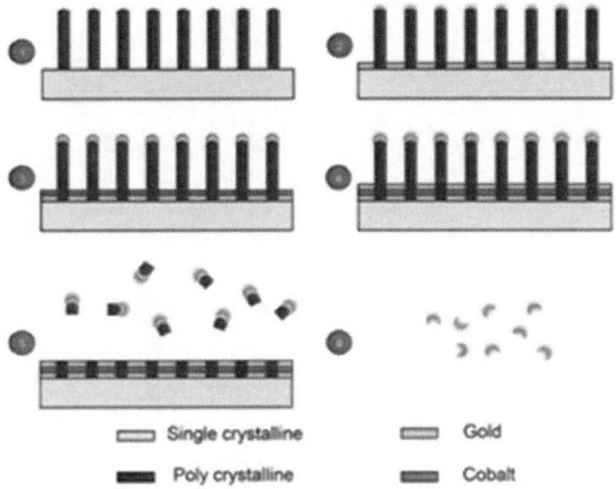

Figure 5. (Reprinted from [100]. Preparation scheme of Au-Co sandwitch structure on silicon pillars: 1. Pillars; 2. Gold deposited on pillars followed by 3. a layer of cobalt and 4. another layer of gold; 5. The pillars are etched away leaving 6. nanowontons structure.

3. FUTURE IMAGING

Among various approaches for multimodal bio-imaging with magnetic NPs, theranostics (therapy and diagnosis) is thought to be the most promising [101]. Liu recently reported the preparation of albumin-based microbubbles incorporating SPIONs and camptothecin as an anti-cancer agent for US-triggered drug release [102]. In addition, gadolinium oxide NPs conjugated with boron compounds were prepared and used for neutron capture therapy of cancers [103]. The *in vivo* functional imaging of a metabolic reaction would be an interesting application of field by using multimodal bio-imaging probes. Photo-uncaging NPs are being studied as a bimodal bio-imaging probe for MRI and fluorescence imaging [104]. Combined CdSe/ZnS QDs and SPIONs with a photoactive linker as well as *in vivo* detection by both MRI and fluorescent imaging after uncaging may be effective for elucidating the excretion fate of uncaged QDs and SPIONs after intravenous and subcutaneous injection. These new approaches suggest that tracing the variation of enzyme-cleaved ligands by multimodal bio-imaging probes could be useful for visualizing *in vivo* metabolic reactions and/or other bio-functions.

CONCLUSION

Over the past ten years, much attention has been devoted to examining various combinations of imaging modalities at the laboratory level. The next stage is to move toward a translational study, with precise validation of the actual worth of multimodal bio-imaging for the detection of diseases at a super-early stage or with extreme accuracy for the efficient treatment of disease, which should improve the quality of life (QOL) of patients. The further theranostic application of magnetic NPs, including for functionalized multimodal bio-imaging and targeted drug delivery, is now growing rapidly. With the further refinement of diagnostic modality hardware, the clinical application of these novel magnetic NPs is expected.

REFERENCES

[1] De la Zerda A; Zavaleta C; Keren S; Vaithilingam S; Bodapati S; Liu Z, et al. Carbon nanotubes as photoacoustic molecular imaging agents in living mice. *Nat Nanotechnol*, 2008 Sep;3(9),557-62.

[2] Wang LV. Ultrasound-mediated biophotonic imaging: a review of acousto-optical tomography and photo-acoustic tomography. *Dis Markers*, 2003;19(2-3),123-38.

[3] Gubin SP, editor. *Magnetic Nanoparticles*. Weinheim: Wiley-VCH; 2009.

[4] Thahn TKN, editor. *Magnetic nanoparticles: From fabrication to clinical applications*. Boca Raton: CRC Press; 2013.

[5] Lu AH; Salabas EL; Schuth F. Magnetic nanoparticles: synthesis, protection, functionalization, and application. *Angew Chem Int Ed Engl*, 2007;46(8),1222-44.

[6] Schladt TD; Schneider K; Schild H; Tremel W. Synthesis and bio-functionalization of magnetic nanoparticles for medical diagnosis and treatment. *Dalton Trans*, 2011 Jun 28;40(24),6315-43.

[7] Chen H; Zhen Z; Todd T; Chu PK; Xie J. Nanoparticles for improving cancer diagnosis. *Mater Sci Eng R Rep*, 2013 Mar;74(3),35-69.

[8] Bazzi R; Flores-Gonzalez MA; Louis C; Lebbou K; Dujardin C; Brenier A, et al. Synthesis and luminescent properties of sub-5-nm lanthanide oxides nanoparticles. *J Lumin*, 2003 May;102,445-50.

[9] Lee Y; Lee J; Bae CJ; Park JG; Noh HJ; Park JH, et al. Large-scale synthesis of uniform and crystalline magnetite nanoparticles using reverse micelles as nanoreactors under reflux conditions. *Adv Funct Mater*, 2005;15(3),503-9.

[10] Jun YW; Lee JH; Cheon J. Chemical design of nanoparticle probes for high-performance magnetic resonance imaging. *Angew Chem Int Ed Engl*, 2008;47(28),5122-35.

[11] Nam J; Won N; Bang J; Jin H; Park J; Jung S, et al. Surface engineering of inorganic nanoparticles for imaging and therapy. *Adv Drug Deliv Rev*, 2013 May;65(5),622-48.

[12] Tu C; Louie AY. Nanoformulations for molecular MRI. *Wiley Interdiscip Rev Nanomed Nanobiotechnol*, 2012 Jul-Aug;4(4),448-57.

[13] Louie A. Multimodality imaging probes: design and challenges. *Chem Rev*, 2010 May 12;110(5),3146-95.

[14] Banerjee SS; Chen D-H. A multifunctional magnetic nanocarrier bearing fluorescent dye for targeted drug delivery by enhanced two-photon triggered release. *Nanotechnology*, 2009;20(18),185103.

[15] Gallagher JJ; Tekoriute R; O'Reilly J-A; Kerskens C; Gun'ko YK; Lynch M. Bimodal magnetic-fluorescent nanostructures for biomedical applications. *J Mater Chem*, 2009;19(24),4081-4.

[16] Maxwell DJ; Bonde J; Hess DA; Hohm SA; Lahey R; Zhou P, et al. Fluorophore-conjugated iron oxide nanoparticle labeling and analysis of engrafting human hematopoietic stem cells. *Stem Cells*, 2008 Feb;26(2),517-24.

[17] Gunn J; Wallen H; Veiseh O; Sun C; Fang C; Cao J, et al. A multimodal targeting nanoparticle for selectively labeling T cells. *Small*, 2008 Jun;4(6),712-5.

[18] Cheng CM; Chu PY; Chuang KH; Roffler SR; Kao CH; Tseng WL, et al. Hapten-derivatized nanoparticle targeting and imaging of gene expression by multimodality imaging systems. *Cancer Gene Ther*, 2009 Jan;16(1),83-90.

[19] Hirata N; Tanabe K; Narita A; Tanaka K; Naka K; Chujo Y, et al. Preparation and fluorescence properties of fluorophore-labeled avidin-biotin system immobilized on Fe_3O_4 nanoparticles through functional indolequinone linker. *Bioorg Med Chem*, 2009 Jun 1;17(11),3775-81.

[20] Min Yeo K; Ji Gao C; Ahn KH; Su Lee I. Superparamagnetic iron oxide nanoparticles with photoswitchable fluorescence. *Chem Commun*, 2008 Oct;14(38),4622-4.

[21] Quarta A; Di Corato R; Manna L; Argentiere S; Cingolani R; Barbarella G, et al. Multifunctional nanostructures based on inorganic nanoparticles and oligothiophenes and their exploitation for cellular studies. *J Am Chem Soc*, 2008 Aug 13;130(32),10545-55.

[22] Trehin R; Figueiredo JL; Pittet MJ; Weissleder R; Josephson L; Mahmood U. Fluorescent nanoparticle uptake for brain tumor visualization. *Neoplasia*, 2006 Apr;8(4),302-11.

[23] Pittet MJ; Swirski FK; Reynolds F; Josephson L; Weissleder R. Labeling of immune cells for *in vivo* imaging using magnetofluorescent nanoparticles. *Nat Protoc*, 2006;1(1),73-9.

[24] Kircher MF; Mahmood U; King RS; Weissleder R; Josephson L. A multimodal nanoparticle for preoperative magnetic resonance imaging and intraoperative optical brain tumor delineation. *Cancer Res*, 2003 Dec 1;63(23),8122-5.

[25] Kelly KA; Allport JR; Tsourkas A; Shinde-Patil VR; Josephson L; Weissleder R. Detection of vascular adhesion molecule-1 expression using a novel multimodal nanoparticle. *Circ Res*, 2005 Feb 18;96(3),327-36.

[26] Sosnovik DE; Schellenberger EA; Nahrendorf M; Novikov MS; Matsui T; Dai G, et al. Magnetic resonance imaging of cardiomyocyte apoptosis with a novel magneto-optical nanoparticle. *Magn Reson Med*, 2005 Sep;54(3),718-24.

[27] Schellenberger EA; Sosnovik D; Weissleder R; Josephson L. Magneto/optical annexin V, a multimodal protein. *Bioconjug Chem*, 2004 Sep-Oct;15(5),1062-7.

[28] Moore A; Medarova Z; Potthast A; Dai G. *In vivo* targeting of underglycosylated MUC-1 tumor antigen using a multimodal imaging probe. *Cancer Res*, 2004;64(5),1821-7.

[29] Montet X; Montet-Abou K; Reynolds F; Weissleder R; Josephson L. Nanoparticle imaging of integrins on tumor cells. *Neoplasia*, 2006 Mar;8(3),214-22.

[30] Medarova Z; Kumar M; Ng SW; Yang J; Barteneva N; Evgenov NV, et al. Multifunctional magnetic nanocarriers for image-tagged siRNA delivery to intact pancreatic islets. *Transplantation*, 2008 Nov 15;86(9),1170-7.

[31] Josephson L; Kircher MF; Mahmood U; Tang Y; Weissleder R. Near-infrared fluorescent nanoparticles as combined MR/optical imaging probes. *Bioconjug Chem*, 2002;13(3),554-60.

[32] Yang L; Mao H; Cao Z; Wang YA; Peng X; Wang X, et al. Molecular imaging of pancreatic cancer in an animal model using targeted multifunctional nanoparticles. *Gastroenterology*, 2009 May;136(5),1514-25.

[33] Veiseh O; Sun C; Gunn J; Kohler N; Gabikian P; Lee D, et al. Optical and MRI multifunctional nanoprobe for targeting gliomas. *Nano Lett*, 2005 Jun;5(6),1003-8.

[34] Lee HY; Li Z; Chen K; Hsu AR; Xu C; Xie J, et al. PET/MRI dual-modality tumor imaging using arginine-glycine-aspartic (RGD)-conjugated radiolabeled iron oxide nanoparticles. *J Nucl Med*, 2008 Aug;49(8),1371-9.

[35] Jarrett BR; Gustafsson B; Kukis DL; Louie AY. Synthesis of ^{64}Cu-labeled magnetic nanoparticles for multimodal imaging. *Bioconjug Chem*, 2008 Jul;19(7),1496-504.

[36] Nahrendorf M; Zhang H; Hembrador S; Panizzi P; Sosnovik DE; Aikawa E, et al. Nanoparticle PET-CT imaging of macrophages in inflammatory atherosclerosis. *Circulation*, 2008 Jan 22;117(3),379-87.

[37] Natarajan A; Gruettner C; Ivkov R; DeNardo GL; Mirick G; Yuan A, et al. NanoFerrite particle based radioimmunonanoparticles: binding affinity and *in vivo* pharmacokinetics. *Bioconjug Chem*, 2008 Jun;19(6),1211-8.

[38] Choi JS; Park JC; Nah H; Woo S; Oh J; Kim KM, et al. A hybrid nanoparticle probe for dual-modality positron emission tomography and magnetic resonance imaging. *Angew Chem Int Ed Engl*, 2008;47(33),6259-62.

[39] Devaraj NK; Keliher EJ; Thurber GM; Nahrendorf M; Weissleder R. ^{18}F labeled nanoparticles for *in vivo* PET-CT imaging. *Bioconjug Chem*, 2009 Feb;20(2),397-401.

[40] Wu SH; Lin YS; Hung Y; Chou YH; Hsu YH; Chang C, et al. Multifunctional mesoporous silica nanoparticles for intracellular labeling and animal magnetic resonance imaging studies. *Chembiochem*, 2008 Jan 4;9(1),53-7.

[41] Lu CW; Hung Y; Hsiao JK; Yao M; Chung TH; Lin YS, et al. Bifunctional magnetic silica nanoparticles for highly efficient human stem cell labeling. *Nano Lett*, 2007 Jan;7(1),149-54.

[42] Makovec D; Čampelj S; Bele M; Maver U; Zorko M; Drofenik M, et al. Nanocomposites containing embedded superparamagnetic iron oxide nanoparticles and rhodamine 6G. *Colloids Surface A*, 2009 Feb 20;334(1-3),74-9.

[43] Ren C; Li J; Liu Q; Ren J; Chen X; Hu Z, et al. Synthesis of organic dye-impregnated silica shell-coated iron oxide nanoparticles by a new method. *Nanoscale Res Lett*, 2008 Dec;3(12),496-501.

[44] Ichiyanagi Y; Moritake S; Taira S; Setou M. Functional magnetic nanoparticles for medical application. *J Magn Magn Mater*, 2007 Mar;310(2),2877-9.

[45] Yoon TJ; Kim JS; Kim BG; Yu KN; Cho MH; Lee JK. Multifunctional nanoparticles possessing a "magnetic motor effect" for drug or gene delivery. *Angew Chem Int Ed Engl*, 2005 Feb 4;44(7),1068-71.

[46] Lee K; Moon H-Y; Park C; Kim OR; Ahn E; Lee SY, et al. Magnetic resonance imaging of macrophage activity in atherosclerotic plaques of apolipoprotein E-deficient mice by using polyethylene glycolated magnetic fluorescent silica-coated nanoparticles. *Curr Appl Phys*, 2009;9(1),S15-S8.

[47] Ma Z; Dosev D; Nichkova M; Dumas RK; Gee SJ; Hammock BD, et al. Synthesis and characterization of multifunctional silica core-shell nanocomposites with magnetic and fluorescent functionalities. *J Magn Magn Mater*, 2009 May 1;321(10),1368-71.

[48] Heitsch AT; Smith DK; Patel RE; Ress D; Korgel BA. Multifunctional particles: magnetic nanocrystals and gold nanorods coated with fluorescent dye-doped silica shells. *J Solid State Chem*, 2008 Jul;181(7),1590-9.

[49] Nagao D; Yokoyama M; Yamauchi N; Matsumoto H; Kobayashi Y; Konno M. Synthesis of highly monodisperse particles composed of a magnetic core and fluorescent shell. *Langmuir*, 2008 Sep 2;24(17),9804-8.

[50] Salgueirino-Maceira V; Correa-Duarte MA; Lopez-Quintela MA; Rivas J. Advanced hybrid nanoparticles. *J Nanosci Nanotechnol*, 2009 Jun;9(6),3684-8.

[51] Santra S; Kaittanis C; Grimm J; Perez JM. Drug/dye-loaded, multifunctional iron oxide nanoparticles for combined targeted cancer therapy and dual optical/magnetic resonance imaging. *Small*, 2009 Aug 17;5(16),1862-8.

[52] Li L; Li H; Chen D; Liu H; Tang F; Zhang Y, et al. Preparation and characterization of quantum dots coated magnetic hollow spheres for magnetic fluorescent multimodal imaging and drug delivery. *J Nanosci Nanotechnol*, 2009 Apr;9(4),2540-5.

[53] Selvan ST; Patra PK; Ang CY; Ying JY. Synthesis of silica-coated semiconductor and magnetic quantum dots and their use in the imaging of live cells. *Angew Chem Int Ed Engl*, 2007;46(14),2448-52.

[54] Du G; Liu Z; Wang D; Xia X; Jia L; Yao K, et al. Characterization of magnetic fluorescence Fe_3O_4/CdSe nanocomposites. *J Nanosci Nanotechnol*, 2009 Feb;9(2),1304-7.

[55] Wang D; He J; Rosenzweig N; Rosenzweig Z. Superparamagnetic Fe_2O_3 beads-CdSe/ZnS quantum dots core-shell nanocomposite particles for cell separation. *Nano Letters*, 2004;4(3),409-13.

[56] Zhang Y; Wang SN; Ma S; Guan JJ; Li D; Zhang XD, et al. Self-assembly multifunctional nanocomposites with Fe_3O_4 magnetic core and CdSe/ZnS quantum dots shell. *J Biomed Mater Res A*, 2008 Jun 1;85(3),840-6.

[57] Dosev D; Nichkova M; Dumas RK; Gee SJ; Hammock BD; Liu K, et al. Magnetic/luminescent core/shell particles synthesized by spray pyrolysis and their application in immunoassays with internal standard. *Nanotechnology*, 2007 Jul;18(5),55102.

[58] He H; Xie MY; Ding Y; Yu XF. Synthesis of Fe_3O_4@LaF_3:Ce,Tb nanocomposites with bright fluorescence and strong magnetism. *Appl Surf Sci*, 2009;255(8),4623-6.

[59] Lu H; Yi G; Zhao S; Chen D; Guo L-H; Cheng J. Synthesis and characterization of multi-functional nanoparticles possessing magnetic, up-conversion fluorescence and bio-affinity properties. *J Mater Chem*, 2004;14(8),1336-41.

[60] Ma ZY; Dosev D; Nichkova M; Gee SJ; Hammock BD; Kennedy IM. Synthesis and bio-functionalization of multifunctional magnetic Fe_3O_4@Y_2O_3:Eu nanocomposites. *J Mater Chem*, 2009;19(27),4695-700.

[61] Cho S-J; Jarrett BR; Louie AY; Kauzlarich SM. Gold-coated iron nanoparticles: a novel magnetic resonance agent for T1 and T2 weighted imaging. *Nanotechnology*, 2006;17(3),640-4.

[62] Larson TA; Bankson J; Aaron J; Sokolov K. Hybrid plasmonic magnetic nanoparticles as molecular specific agents for MRI/optical imaging and photothermal therapy of cancer cells. *Nanotechnology*, 2007;18(32),325101.

[63] Seo WS; Lee JH; Sun X; Suzuki Y; Mann D; Liu Z, et al. FeCo/graphitic-shell nanocrystals as advanced magnetic-resonance-imaging and near-infrared agents. *Nat Mater*, 2006 Dec;5(12),971-6.

[64] Soenen SJ; Vercauteren D; Braeckmans K; Noppe W; De Smedt S; De Cuyper M. Stable long-term intracellular labelling with fluorescently tagged cationic magnetoliposomes. *Chembiochem*, 2009 Jan 26;10(2),257-67.

[65] Zhang Y; Das GK; Xu R; Yang Tan TT. Tb-doped iron oxide: bifunctional fluorescent and magnetic nanocrystals. *J Mater Chem*, 2009;19(22),3696-703.

[66] Sandiford L; Phinikaridou A; Protti A; Meszaros LK; Cui XJ; Yan Y, et al. Bisphosphonate-anchored PEGylation and radiolabeling of superparamagnetic iron oxide: long-circulating nanoparticles for *in vivo* multimodal (T1 MRI-SPECT) imaging. *ACS Nano*, 2013 Jan;7(1),500-12.

[67] Lauterbur PC; Mendonca-Dias M; Rudin A. Augmentation of tissue water proton spin-lattice relaxation rates by *in vivo* addition of paramagnetic ions. In: Dutton PL, Leigh JS, Scarpa A, editors. *Frontiers of biological energetics*. 1. New York: Academic Press; 1978; p. 752-9.

[68] Pan D; Caruthers SD; Senpan A; Schmieder AH; Wickline SA; Lanza GM. Revisiting an old friend: manganese-based MRI contrast agents. *Wiley Interdiscip Rev Nanomed Nanobiotechnol*, 2011 Sep 21;3(April),162-73.

[69] Na HB; Lee JH; An K; Park YI; Park M; Lee IS, et al. Development of a T1 contrast agent for magnetic resonance imaging using MnO nanoparticles. *Angew Chem Int Ed Engl*, 2007;46(28),5397-401.

[70] Huang J; Xie J; Chen K; Bu L; Lee S; Cheng Z, et al. HSA coated MnO nanoparticles with prominent MRI contrast for tumor imaging. *Chem Commun*, 2010 Sep 28;46(36),6684-6.

[71] Shin J; Anisur RM; Ko MK; Im GH; Lee JH; Lee IS. Hollow manganese oxide nanoparticles as multifunctional agents for magnetic resonance imaging and drug delivery. *Angew Chem Int Ed Engl*, 2009;48(2),321-4.

[72] Melnikov OV; Gorbenko OY; Markelova MN; Kaul AR; Atsarkin VA; Demidov VV, et al. Ag-doped manganite nanoparticles: new materials for temperature-controlled medical hyperthermia. *J Biomed Mater Res A*, 2009 Dec 15;91(4),1048-55.

[73] Uskokovic V; Kosak A; Drofenik M; Drofenik M. Preparation of silica-coated lanthanum-strontium manganite particles with designable curie point, for application in hyperthermia treatments. *Int J Appl Ceram Technol*, 2006;3(2),134-43.

[74] Ken-Tye Y. Mn-doped near-infrared quantum dots as multimodal targeted probes for pancreatic cancer imaging. *Nanotechnology*, 2009;20(1),015102.

[75] Lin B; Yao X; Zhu Y; Shen J; Yang X; Jiang H, et al. Multifunctional manganese-doped core–shell quantum dots for magnetic resonance and fluorescence imaging of cancer cells. *New J Chem*, 2013;37(10),3076-83.

[76] Choi SH; Na HB; Park YI; An K; Kwon SG; Jang Y, et al. Simple and generalized synthesis of oxide-metal heterostructured nanoparticles and their applications in multimodal biomedical probes. *J Am Chem Soc*, 2008 Nov 19;130(46),15573-80.

[77] Ha TL; Kim HJ; Shin J; Im GH; Lee JW; Heo H, et al. Development of target-specific multimodality imaging agent by using hollow manganese oxide nanoparticles as a platform. *Chem Commun*, 2011 Aug 28;47(32),9176-8.

[78] Bridot JL; Faure AC; Laurent S; Riviere C; Billotey C; Hiba B, et al. Hybrid gadolinium oxide nanoparticles: multimodal contrast agents for *in vivo* imaging. *J Am Chem Soc*, 2007 Apr 25;129(16),5076-84.

[79] Fortin M-A; Petoral Jr RM; Söderlind F; Klasson A; Engström M; Veres T, et al. Polyethylene glycol-covered ultra-small Gd_2O_3 nanoparticles for positive contrast at 1.5 T magnetic resonance clinical scanning. *Nanotechnology*, 2007;18(39),395501.

[80] Evanics F; Diamente PR; van Veggel FCJM; Stanisz GJ; Prosser RS. Water-soluble $GdF_3$and GdF_3/LaF_3 nanoparticles: physical characterization and NMR relaxation properties. *Chem Mater*, 2006;18(10),2499-505.

[81] Sharma P; Brown SC; Bengtsson N; Zhang Q; Walter GA; Grobmyer SR, et al. Gold-speckled multimodal nanoparticles for noninvasive bioimaging. *Chem Mater*, 2008 Oct 14;20(19),6087-94.

[82] Liu L; Ding H; Yong K-T; Roy I; Law W-C; Kopwitthaya A, et al. Application of gold nanorods for plasmonic and magnetic imaging of cancer cells. *Plasmonics*, 2010;6(1),105-12.

[83] Paik T; Gordon TR; Prantner AM; Yun H; Murray CB. Designing tripodal and triangular gadolinium oxide nanoplates and self-assembled nanofibrils as potential multimodal bioimaging probes. *ACS Nano*, 2013 Mar;7(3),2850-9.

[84] Das GK; Heng BC; Ng S-C; White T; Loo JSC; D'Silva L, et al. Gadolinium oxide ultranarrow nanorods as multimodal contrast agents

for optical and magnetic resonance imaging. *Langmuir*, 2010;26,8959-65.

[85] Petoral RM; Söderlind F; Klasson A; Suska A; Fortin MA; Abrikossova N, et al. Synthesis and characterization of Tb^{3+}-doped Gd_2O_3 nanocrystals: a bifunctional material with combined fluorescent labeling and MRI contrast agent properties. *J Phys Chem C*, 2009;113(17),6913-20.

[86] Sitharaman B; Jacobson BD; Wadghiri YZ; Bryant H; Frank J. The magnetic, relaxometric, and optical properties of gadolinium-catalyzed single walled carbon nanotubes. *J Appl Phys*, 2013 Apr 7;113(13),134308.

[87] Zheng ZC; Daniel A; Yu W; Weber B; Ling J; Muller AHE. Rare-earth metal cations incorporated silica hybrid nanoparticles templated by cylindrical polymer brushes. *Chem Mater*, 2013 Nov 26;25(22),4585-94.

[88] Zhu D; Liu F; Ma L; Liu D; Wang Z. Nanoparticle-based systems for T1-weighted magnetic resonance imaging contrast agents. *Int J Mol Sci*, 2013 May;14(5),10591-607.

[89] Liu C; Gao Z; Zeng J; Hou Y; Fang F; Li Y, et al. Magnetic/upconversion fluorescent $NaGdF_4$:Yb,Er nanoparticle-based dual-modal molecular probes for imaging tiny tumors *in vivo*. *ACS Nano*, 2013;7(8),7227-40.

[90] Kimura Y; Kamisugi R; Narazaki M; Matsuda T; Tabata Y; Toshimitsu A, et al. Size-controlled and biocompatible Gd_2O_3 nanoparticles for dual photoacoustic and MR imaging. *Adv Healthcare Mater*, 2012 Sep;1(5),657-60.

[91] Sitharaman B; Kissell KR; Hartman KB; Tran LA; Baikalov A; Rusakova I, et al. Superparamagnetic gadonanotubes are high-performance MRI contrast agents. *Chem Commun*, 2005 Aug 21(31),3915-7.

[92] Maeda H. Tumor-selective delivery of macromolecular drugs via the EPR effect: background and future prospects. *Bioconjug Chem*, 2010 May 19;21(5),797-802.

[93] Guay-Begin AA; Chevallier P; Faucher L; Turgeon S; Fortin MA. Surface modification of gadolinium oxide thin films and nanoparticles using poly(ethylene glycol)-phosphate. *Langmuir*, 2012 Jan 10;28(1),774-82.

[94] Soderlind F; Pedersen H; Petoral RM, Jr.; Kall PO; Uvdal K. Synthesis and characterisation of Gd_2O_3 nanocrystals functionalised by organic acids. *J Colloid Interface Sci*, 2005 Aug 1;288(1),140-8.

[95] Kattel K; Park JY; Xu W; Kim HG; Lee EJ; Bony BA, et al. A facile synthesis, *in vitro* and *in vivo* MR studies of D-glucuronic acid-coated ultrasmall Ln_2O_3 (Ln = Eu, Gd, Dy, Ho, and Er) nanoparticles as a new potential MRI contrast agent. *ACS Appl Mater Interfaces*, 2011 Sep;3(9),3325-34.
[96] Bhakta G; Sharma RK; Gupta N; Cool S; Nurcombe V; Maitra A. Multifunctional silica nanoparticles with potentials of imaging and gene delivery. *Nanomedicine*, 2011 Aug;7(4),472-9.
[97] Mikhaylova M; Stasinopoulos I; Kato Y; Artemov D; Bhujwalla ZM. Imaging of cationic multifunctional liposome-mediated delivery of COX-2 siRNA. *Cancer Gene Ther*, 2009 Mar;16(3),217-26.
[98] Sciallero C; Grishenkov D; Kothapalli SV; Oddo L; Trucco A. Acoustic characterization and contrast imaging of microbubbles encapsulated by polymeric shells coated or filled with magnetic nanoparticles. *J Acoust Soc Am*, 2013 Nov;134(5),3918-30.
[99] Li A; Zheng YY; Yu JH; Wang ZG; Yang Y; Wu W, et al. Superparamagnetic perfluorooctylbromide nanoparticles as a multimodal contrast agent for US, MR, and CT imaging. *Acta Radiologica*, 2013 Apr;54(3),278-83.
[100] Bouchard LS; Anwar MS; Liu GL; Hann B; Xie ZH; Gray JW, et al. Picomolar sensitivity MRI and photoacoustic imaging of cobalt nanoparticles. *Proc Natl Acad Sci USA*, 2009 Mar 17;106(11),4085-9.
[101] Ahmed N; Fessi H; Elaissari A. Theranostic applications of nanoparticles in cancer. *Drug Discov Today*, 2012 Sep;17(17-18),928-34.
[102] Liu TY; Wu MY; Lin MH; Yang FY. A novel ultrasound-triggered drug vehicle with multimodal imaging functionality. *Acta Biomater*, 2013 Mar;9(3),5453-63.
[103] Kim SM; Chae MK; Yim MS; Jeong IH; Cho J; Lee C, et al. Hybrid PET/MR imaging of tumors using an oleanolic acid-conjugated nanoparticle. *Biomaterials*, 2013 Nov;34(33),8114-21.
[104] Shibu ES; Ono K; Sugino S; Nishioka A; Yasuda A; Shigeri Y, et al. Photouncaging nanoparticles for MRI and fluorescence imaging *in vitro* and *in vivo*. *ACS Nano*, 2013 Nov 26;7(11),9851-9.

In: Magnetic Nanoparticles
Editor: Nora P. Sabbas
ISBN: 978-1-63117-434-6

Chapter 4

Synthesis, Size Separation, Characterization and Surface Engineering of Superparamagnetic Iron Oxide Nanoparticles for Biomedical Applications

Li Zhao and Naoki Komatsu [*]
Department of Chemistry, Shiga University of Medical Science, Seta, Otsu, Japan

Abstract

Superparamagnetic iron oxide nanoparticle (SPION) has attracted a great deal of attention in the fields of medicine and biology due to its non-toxicity, large specific surface area, amenability to various surface fuctionalization, and unique magnetic properties. For biomedical applications, we synthesized two kinds of SPIONs by co-precipitation and polyol methods, designated as *C*- and *P*-SPIONs, respectively. To increase their solubility and stability in an aqueous solution, both of the SPIONs were functionalized with hyperbranched polyglycerol (PG) through ring-opening polymerization of glycidol. Scanning transmission electron microscopy (STEM) and dynamic light scattering (DLS) measurements revealed that the *P*-SPION provided individual

[*] Corresponding Author: Naoki Komatsu, E-mail address: nkomatsu@belle.shiga-med.ac.jp.

nanoparticles without aggregation after the PG fuctionalization. This *P*-SPION-PG was highly soluble not only in pure water (> 40 mg/mL), but also in a phosphate buffer solution (> 25 mg/mL). Such high solubility enabled separation of *P*-SPION-PG according to their size by size exclusion chromatography (SEC). The size-separated *P*-SPION-PG shows a gradual increase in transverse relaxivity (r_2) with increasing particle size. The hydroxyl group in PG also serves as a scaffold for further surface functionalization. For targeted cell labeling, we functionalized *P*-SPION-PG through multi-step organic transformations (-OH → -OTs (tosylate) → $-N_3$ → -RGD) including click chemistry as a key step to impart targeting specificity by immobilization of cyclic RGD peptide on the surface. The targeting effect was demonstrated by the cell experiments; *P*-SPION-PG-RGD was taken up by the cells overexpressing $\alpha_v\beta_3$-integrin such as U87MG and A549. In addition, near-infrared (NIR) fluorescence dye (IRdye 800CW) was immobilized on the surface of *P*-SPION-PG through amide linkage. The resulting *P*-SPION-PG-IRdye showed good solubility in phosphate buffered saline (PBS) and strong fluorescence in NIR region, which may find application in optical/MR bimodal imaging.

Keywords: Iron Oxide, Nanoparticle, Polyglycerol, Functionalization, Targeting; Imaging

1. INTRODUCTION

In recent years, superparamagnetic iron oxide nanoparticle (SPION) has attracted significant attention in a field of nanomedicine due to its non-toxicity, good biocompatibility, high specific surface area, and unique magnetic properties. SPION has been successfully used as an MR imaging probe for diagnosis as well as a drug carrier and a hyperthermic agent for therapy [1-8]. For these applications, appropriate surface coating is indispensible to provide SPION with various functions such as aqueous solubility, active targeting, and therapeutic ability. In order to achieve high solubility and stability in a physiological medium, most of the commercial SPIONs are coated with dextran and its derivatives.[5] Although dextran-coating imparts good aqueous solubility to SPION, this results in increase of the particle size and broadening of the size distribution compared to the size and size distribution of the individual SPION. Another surface modification technique frequently employed is immobilization of polyethylene glycol (PEG) on the surface of SPION through anchoring ligands (dopamine, phosphonic acids, and trimethoxysilane etc.) with high affinity to SPION [9]. Although PEG has been considered to be the most effective surface functionality, especially, due to the stealth nature, recent studies reported its disadvantages, designated as "PEG dilemma", such as immune response and interference with cellular uptake [10].

Polyglycerol (PG) grafting through ring-opening polymerization of glycidol has been developed as a facile and efficient method to endow a verity of nanoparticles with extremely high aqueous solubility. For example, PG grafted nanodiamond (ND-PG) showed several hundred times larger solubility than PEG-modified ND [11, 12]. Since hydroxyl groups on the hydrophilic SPION were proven to serve as scaffolds for multi-step covalent transformations on the surface [13], we can use directly the hydroxyl groups to initiate the ring-opening polymerization. In addition, a large number of hydroxyl groups in PG offer the scaffold for further functionalization, adding more functions including targeting moieties and fluorescent tags.

In our recent paper [14], we have reported on the preparation of SPION by polyol method, direct PG grafting on the surface, and further conjugation with RGD peptide for targeted cellular labeling. We have also realized chromatographic size separation of the PG-functionalized SPION (SPION-PG) and confirmed the relationship between particle size and transverse relaxivity. In this chapter, more details and further extensions regarding synthesis, functionalization, and characterization of SPION and SPION-PG were incorporated, including preparation of SPION by conventional co-precipitation method, following PG grafting of SPION, spectroscopic, gravimetric and morphological analyses, and conjugation of SPION-PG with near-infrared (NIR) fluorescent dye for optical/MR bimodal imaging.

2. Results and Discussion

2.1 Synthesis and Characterization of SPION and SPION-PG

Co-precipitation and polyol methods were employed to synthesize hydrophilic SPIONs designated as *C*- and *P*-SPIONs, respectively (Scheme 1). In the co-precipitation method, ferric and ferrous chlorides with a molar ratio of 2 : 1 were used as iron sources. Fe_3O_4 nanoparticles (*C*-SPION) were produced by adding concentrated ammonia into the aqueous solution of the above salts under N_2 atmosphere [15]. The chemical reaction formula for *C*-SPION synthesis is shown in eq. 1.

$$2Fe^{3+} + Fe^{2+} + 8OH^- \longrightarrow Fe_3O_4 + 4H_2O \quad (1)$$

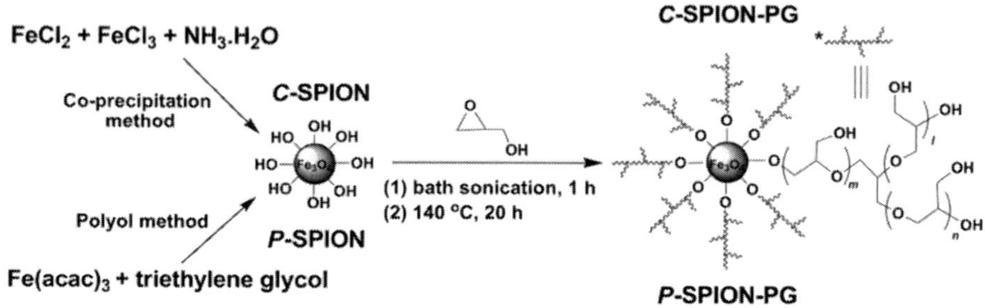

Scheme 1. Synthesis of *C*- and *P*-SPIONs by co-precipitation and polyol methods, and *C*- and *P*-SPION-PGs through ring-opening polymerization of glycidol.

In the polyol method, Fe_3O_4 nanoparticles (*P*-SPION) were synthesized by thermal decomposition of iron (III) acetylacetonate in triethylene glycol (TREG) at high temperature according to the reported procedure [16, 17].

In order to increase the aqueous solubility of SPION in view of biomedical applications, both *C*- and *P*-SPIONs were grafted with PG through ring-opening polymerization of glycidol (Scheme 1). The SPION was well dispersed in glycidol with the aid of bath sonication. The resulting suspension of SPION in glycidol was stirred at 140 °C for 20 h without any additives such as acid and base for the ring-opening polymerization [11, 18]. This is because we found that PG functionalization of ND at high temperature under neutral conditions is better than that under acidic and basic ones to make the ND-PG more soluble in a physiological medium [11].

The *C*- and *P*-SPIONs, and *C*- and *P*-SPION-PGs were characterized by FTIR and thermogravimetric analysis (TGA). Hydroxyl groups on the surface of SPIONs were confirmed by the absorption band at 3400 cm^{-1} corresponding to O–H stretching in IR spectra (Figure 1). The absorption bands of C–H at 2900 cm^{-1} and C–O–C at 1100 cm^{-1} are observed in IR spectrum of *P*-SPION (Figure 1b) more clearly than those of *C*-SPION (Figure 1a), indicating that TREG is attached on the surface of *P*-SPION. The PG grafting on the SPIONs was confirmed by large increase in the absorptions corresponding to C–H and C–O–C stretchings. The TGA profile of *P*-SPION (Figure 2b) showed a weight reduction of 8.9 % at 450 °C, while almost no weight loss was observed in *C*-SPION at the same temperature (Figure 2a). This phenomena support the existence of TREG on the surface of *P*-SPION. The weight percentages of the grafted PGs on the *C*- and *P*-SPION-PGs were calculated to be 50 % and 54 %, respectively (Figure 2).

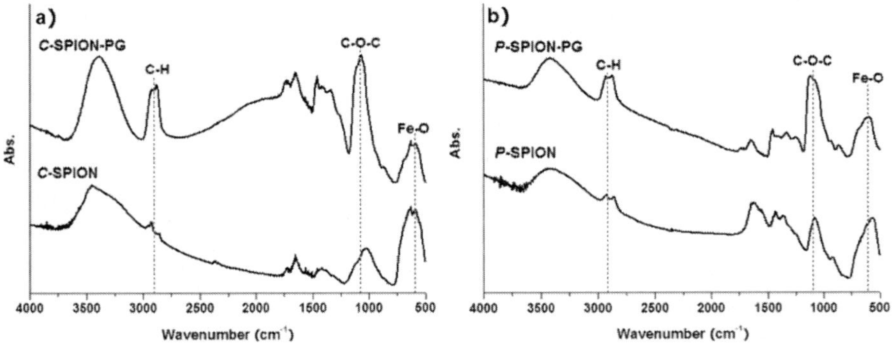

Figure 1. FTIR spectra of a) *C*-SPION and *C*-SPION-PG, and b) *P*-SPION and *P*-SPION-PG.

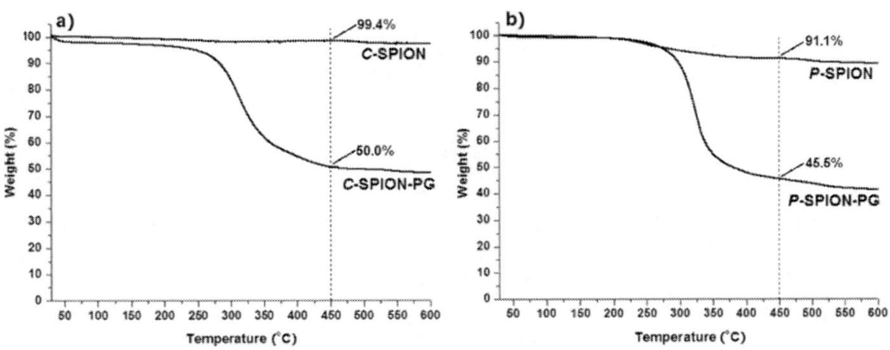

Figure 2. TGA profiles of a) *C*-SPION and *C*-SPION-PG, and b) *P*-SPION and *P*-SPION-PG under nitrogen.

The high hydrophilicity of the grafted PG endowed SPIONs with extremely high aqueous solubility. The solubility of both SPION-PGs is > 40 mg/mL in pure water and > 25 mg/mL in phosphate buffered saline (PBS). However, the aqueous solutions of SPION-PGs showed different stability. After standing at 4 °C for 3 months, clear precipitation was observed in the water solution of *C*-SPION-PG (Figure 3a). In contrast, the water solution of *P*-SPION-PG maintained homogeneous; no noticeable precipitates were observed after 3 months (Figure 3b). In addition, strong hydrophilicity and superparamagnetism of *P*-SPION-PG are simultaneously demonstrated in

Figure 3c. When a very thick aqueous solution of *P*-SPION-PG (40 mg/mL) was subjected to an external magnetic field, the hydrosol containing the magnetic nanoparticles and the aqueous medium were attracted to the magnet, which can be realized only by the highly soluble material with high magneticity.

Figure 3. Water dispersions (1.0 mg/mL) of a) *C*-SPION-PG and b) *P*-SPION-PG after standing for 3 months, and c) an aqueous solution of *P*-SPION-PG (40 mg/mL) in response to a permanent magnet.

To investigate particle size and morphology, SPIONs and SPION-PGs were characterized by scanning transmission electron microscopy (STEM) and dynamic light scattering (DLS). As shown in the STEM images, the primary particles of the *C*- and *P*-SPIONs exhibited similar sizes of *ca.* 9 nm (Figure 4a and 4c, respectively). *C*-SPION is prone to form large clusters with the size of > 100 nm, whereas most of the *P*-SPION exist individually on the substrate probably due to the isolation by the hydrophilic layers supported by TREG layer on the surface of the *P*-SPION.

The difference in the propensity for clustering is reflected by that in the hydrodynamic sizes of *C*- and *P*-SPIONs shown in Table 1. After the PG grafting, *C*-SPION-PG still formed clusters with reduced size; only a few individual nanoparticles were observed in Figure 4b. In contrast, all *P*-SPION-PG particles were found to be individually dispersed on the substrate. As compared with *P*-SPION (Figure 4c), interparticle distances of the *P*-SPION-PG clearly increase in the STEM image (Figure 4d).

This is attributable to thick PG layer grafted on each *P*-SPION particle. The mean hydrodynamic diameters of *C*- and *P*-SPION-PGs in water were determined to be 61.1 ± 22.2 nm and 25.1 ± 5.1 nm by DLS in number distribution, respectively (Table 1).

Based on the difference in the core and hydrodynamic sizes, the thickness of PG layer on the *P*-SPION-PG is calculated to be *ca.* 8 nm.

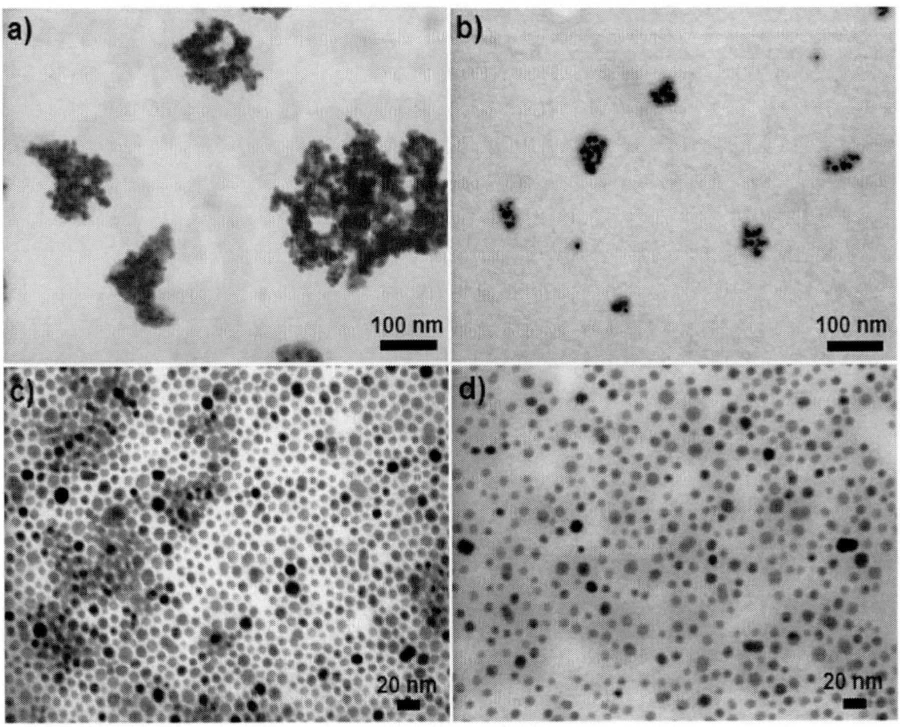

Figure 4. STEM images of a) *C*-SPION, b) *C*-SPION-PG, c) *P*-SPION, and d) *P*-SPION-PG.

Table 1. Particle size of *C*- and *P*-SPIONs, and *C*- and *P*-SPION-PGs, and magnetic property of *P*-SPION-PG before and after chromatographic separation

Sample	Retention time [min]	Core size [nm][a]	Hydrodynamic size [nm][b]	r_2 [mM^{-1} s^{-1}]
C-SPION	-	9.6 ± 2.6	112.0[b]	-
C-SPION-PG	-	9.4 ± 2.5	61.1[b]	-
P-SPION	-	8.8 ± 2.2	67.4[b]	-
P-SPION-PG	21.0-29.0	8.8 ± 2.3	25.1[b], 26.6[c]	86.30
fraction 1	21.0-24.0	10.2 ± 2.7	30.4[c]	91.97
fraction 2	24.0-26.0	9.1 ± 1.9	25.7[c]	86.91
fraction 3	26.0-29.0	7.8 ± 1.7	20.7[c]	77.91

[a] Average core size of SPION determined by more than 200 particles in the STEM images (Figures 4a – 4d and 5b – 5d). [b] Median size determined by DLS in Milli-Q water, unless otherwise noted. [c] In buffer.

2.2. Chromatographic Size Separation of Highly Soluble *P*-SPION-PG and Size-Dependant Transverse Relaxivity

High aqueous solubility of *P*-SPION-PG enables chromatographic separation as in the case of ND-PG.[11] A phosphate buffer (pH 7.0, 20 mM) containing Na_2SO_4 (100 mM) was used as a mobile phase (flow rate: 1.0 mL/min). The elution was monitored with absorption at 254 nm. The typical elution profiles of *P*-SPION-PG and free PG are shown in Figure 5a. The chromatographic peak at the retention time of 21 – 29 min corresponds to the elution of *P*-SPION-PG. A small peak at 35 min is attributed to free PG included in *P*-SPION-PG as an impurity, because the retention time is in accordance with that of the authentic free PG [11]. Three fractions were collected at the retention time of 21 – 24, 24 – 26, and 26 – 29 min from the beginning to the end of *P*-SPION-PG elution. These fractions were analyzed by DLS to determine the hydrodynamic diameters of *P*-SPION-PG. The core size was determined by STEM after desalinization through dialysis against Milli-Q water. The results are summarized in Table 1. As the fraction proceeds from 1 to 3, the core and hydrodynamic diameters decrease from 10.2 to 7.8 nm (Figure 5b-d) and from 30.4 to 20.7 nm (Table 1), respectively. Post-synthesis SEC offers more precise way to control the particle size.

In order to correlate the size parameters of the *P*-SPION-PG with the magnetic properties, MRI transverse relaxivity (r_2) were determined by the slope of the linear approximation based on the following equation (eq. 2):

$$1/T_2 = 1/T^0{}_2 + r_2\,[\mathrm{Fe}] \qquad (2)$$

where T_2 and $T^0{}_2$ are the observed transverse relaxation times of *P*-SPION-PG and pure water, respectively, and [Fe] is the iron concentration of the solutions. The r_2 of as-synthesized *P*-SPION-PG is found to be 86.30 Fe mM^{-1} s^{-1} (Table 1), which is similar to that of the SPION reported by Cai *et al.* (particle size: 8 ± 1.1 nm).[17] The r_2 decreases from fraction 1 to 3 shown in Table 1 can be ascribed to size decrease of the core in *P*-SPION-PG, because the large magnetic particles possess higher magnetic moments and hence distort the magnetic field more efficiently to enhance r_2 [19, 21].

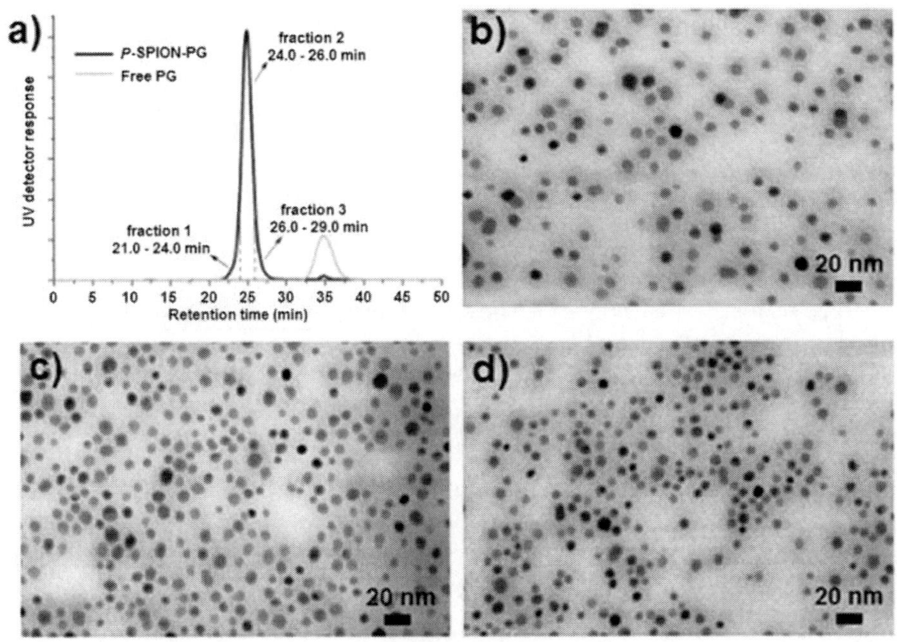

Figure 5. a) Chromatograms of the elution of *P*-SPION-PG and free PG, and STEM images of b) fraction 1, c) fraction 2 and d) fraction 3 after chromatographic separation of *P*-SPION-PG.

2.3. Further Surface Functionalization of *P*-SPION-PG for Biomedical Applications

2.3.1. Conjugation with RGD Peptide for Targeted Cellular Labeling

The PG on the surface of *P*-SPION not only enhances hydrophilicity to provide SPION with high aqueous solubility and stability as described above, but also affords extensibility to further functionalization to add more functions and properties to *P*-SPION-PG. Since PG is also reported to be biocompatible [22] and resistant to protein adsorption [23], PG is considered as one of the ideal surface functionality for nanoparticle in view of biomedical applications. Therefore, we immobilized targeting functionality at the hydroxyl groups in the PG layer through stepwise organic transformation shown in Scheme 2. The resulting *P*-SPION-PG-RGD having RGD peptide as targeting moiety was applied to cells to confirm the targeting efficiency [24].

To introduce cyclic RGD peptide to target the $\alpha_v\beta_3$-integrin expressed in some kinds of cancer cell, the hydroxyl groups (–OH) in the *P*-SPION-PG were converted to tosylate (*P*-SPION-PG-OTs), which is a good leaving group frequently used in organic synthesis, and subsequently to azido (*P*-SPION-PG-N_3) through nucleophilic substitution of the tosylate with sodium azide [25]. The *P*-SPION-PG-N_3 was reacted with cyclic RGD peptide having acetylene terminus through click chemistry [19, 26, 27], giving *P*-SPION-PG-RGD (Scheme 2). These organic transformations were confirmed qualitatively by FTIR [14].

The resulting *P*-SPION-PG-RGD exhibited good solubility and stability in PBS as shown in Figure 6a and was applied to cell experiment to confirm the $\alpha_v\beta_3$-integrin targeted cell labeling. *P*-SPION-PG and *P*-SPION-PG-RGD were incubated with following three kinds of cancer cells; HeLa, U87MG, and A549. The uptake of the nanoparticles was confirmed histologically by Prussian blue staining [28]. No noticeable color change is observed in the cells treated with *P*-SPION-PG (upper pictures in Figure 6b), indicating that *P*-SPION-PG cannot be internalized into the cells through the cell membrane. In contrast, U87MG and A549 treated by *P*-SPION-PG-RGD were stained blue, while HeLa was not stained (lower pictures in Figure 6b). Since cyclic RGD peptide is known to have affinity to $\alpha_v\beta_3$-integrin [29, 30], a degree of blue staining may depend on number of the integrin expressed in the cells. U87MG and A549 are well-known to overexpress the integrin [31, 32], facilitating the internalization of *P*-SPION-PG-RGD into the cells. On the other hand, HeLa is concluded to have much less number of the integrin than U87MG and A549 [33]. In the stained cells shown in Figure 6b, sporadic blue spots were detected in the perinuclear cytoplasmic vesicles of U87MG and A549 cells. These results suggest that *P*-SPION-PG-RGD was incorporated into the cells through receptor-mediated endocytosis based on the affinity of cyclic RGD peptide to the integrin [29, 30, 33]. For any target, the targeting peptide can be identified by phage display method and immobilized on the PG layer through click chemistry in place of the RGD peptide [34].

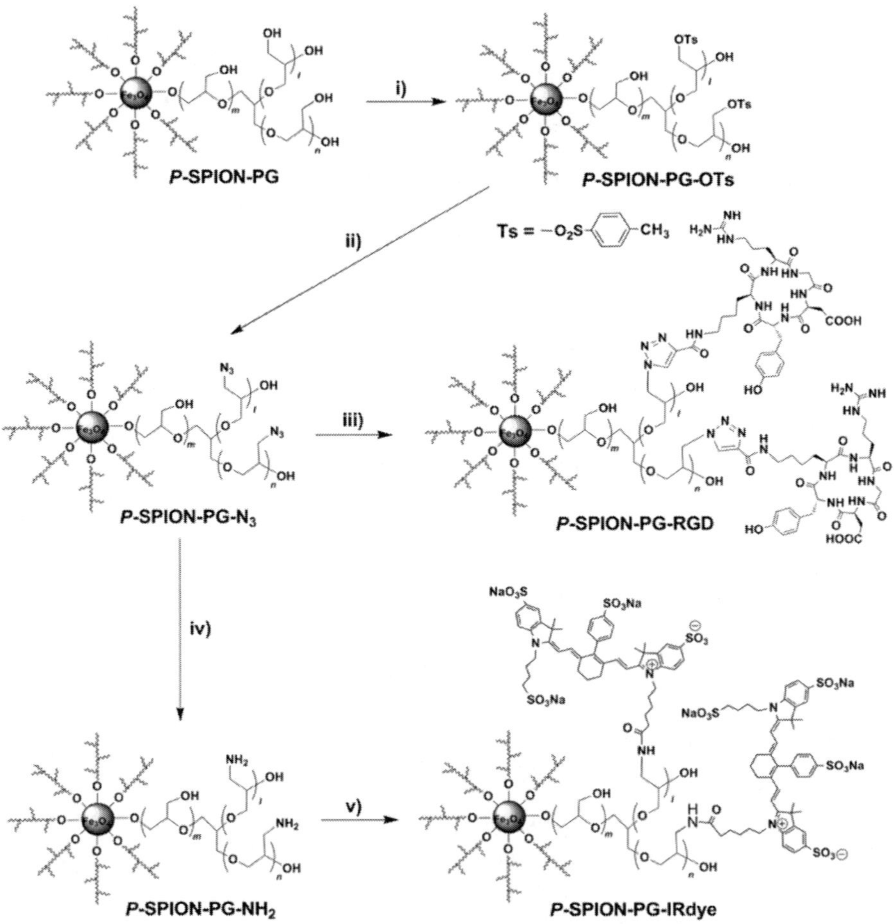

Scheme 2. Synthesis of *P*-SPION-PG-RGD and *P*-SPION-PG-IRdye from *P*-SPION-PG through multi-step organic transformations. i) *p*-TsCl, NaOH, 0 °C – r.t., overnight; ii) NaN_3, 90 °C, 6 h; iii) RGD propargyl amide, copper(II) sulfate pentahydrate, sodium ascorbate, r. t., 36 h; iv) triphenylphosphine, DMF, ammonia, r.t., 30 h; v) IRdye 800CW NHS ester, pH 8.0, r.t., 2 h.

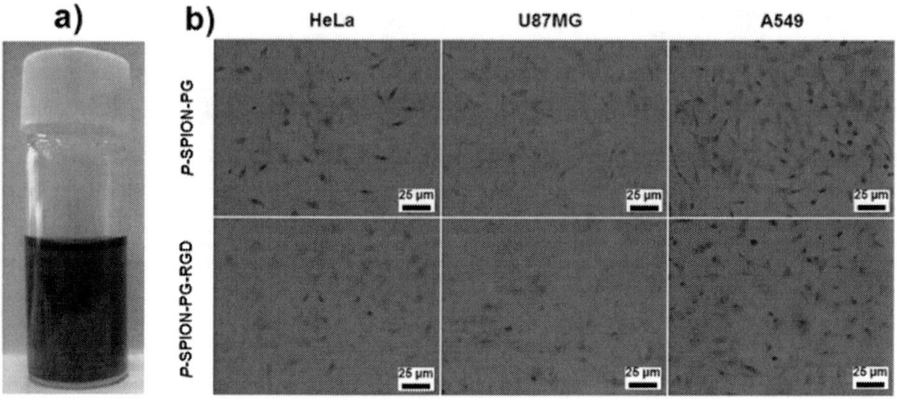

Reprinted with permission [14]. Copyright 2012 Wiley.

Figure 6. a) Photograph of *P*-SPION-PG-RGD in PBS (1 mg/mL); b) Prussian blue stained images of cancer cells treated with *P*-SPION-PG (upper) and *P*-SPION-PG-RGD (lower).

2.3.2. Conjugation with IRdye 800CW for Optical/MR Bimodal Imaging

Multimodal imaging is highly desirable for more accurate diagnosis, because it can provide complementary information from multiple imaging modalities [35]. To achieve optical/MR bimodal imaging, NIR fluorescent dye (IRdye 800CW, LI-COR, Inc.) was immobilized on the surface of *P*-SPION-PG through covalent linkage as shown in Scheme 2. To generate amino groups for conjugation, *P*-SPION-PG-N_3 was reduced with triphenylphosphine to give amine (*P*-SPION-PG-NH_2). The NIR fluorescent dye was connected through amide linkage by reacting *P*-SPION-PG-NH_2 with IRdye 800CW NHS ester, yielding *P*-SPION-PG-IRdye.

The organic transformations and conjugation were confirmed by FTIR (Figure 7). The conversion of azido groups to amino ones was verified by the disappearance of azido absorption at 2100 cm^{-1} (Figure 7a) and the appearance of the absorption bands at 1598 cm^{-1}, which corresponds to the N-H bending vibration of primary amines (Figure 7b). After the conjugation with NIR fluorescent tag, the absorption band of amine was partially consumed and a new band at 1524 cm^{-1} attributed to conjugated C=C stretching of the dye was detected (Figure 7c).

The photophysical properties of *P*-SPION-PG-IRdye were characterized by absorption and photoluminescence spectroscopies, and NIR fluorescence imaging. As demonstrated in the inset of Figure 8a, the water dispersion of *P*-SPION-PG-IRdye shows distinct green color due to the dark green color of the

IRdye 800CW. Absorption spectrum of *P*-SPION-PG-IRdye revealed two absorption bands peaked at 640 and 790 nm (Figure 8a). The latter is in consistence with the absorption band of IRdye 800CW with slight shift. Although the origin of the absorption band at 640 nm is not clear yet, *P*-SPION-PG-IRdye showed fluorescence peak at 800 nm, which coincides with the fluorescence of the IRdye 800CW (Figure 8b). As expected, intense NIR emission was observed in the fluorescence image of the aqueous solution of *P*-SPION-PG-IRdye at the excitation of 785 nm as shown in the inset of Figure 8. The strong NIR emission of *P*-SPION-PG-IRdye together with the aforementioned MRI relaxivities of *P*-SPION-PG suggests that *P*-SPION-PG-IRdye should be a promising contrast agent for in vivo optical/MR bimodal imaging.

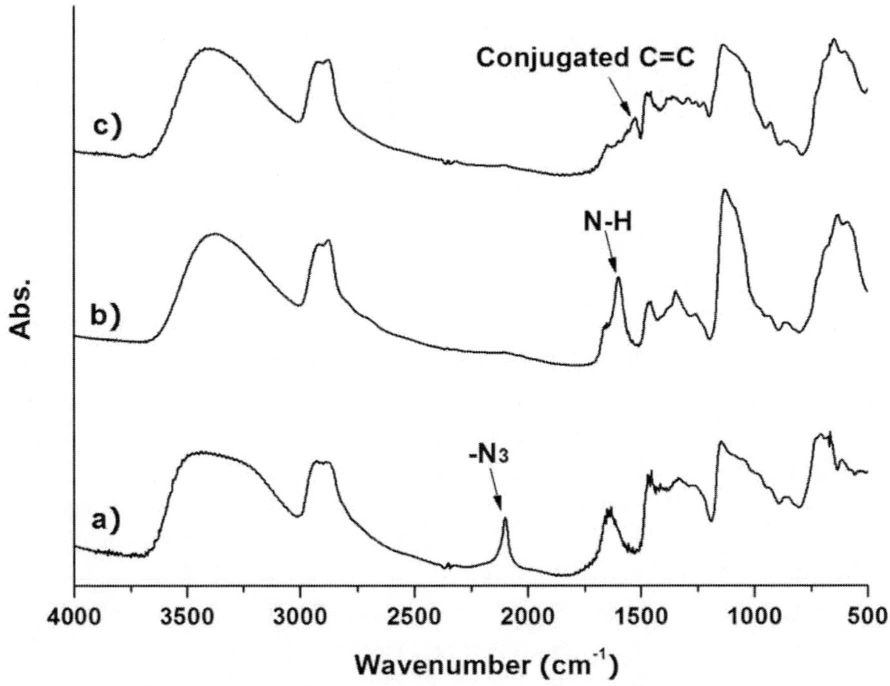

Figure 7. FTIR spectra of a) *P*-SPION-PG-N_3, b) *P*-SPION-PG-NH_2 and b) *P*-SPION-PG-IRdye. Arrows indicate new absorption band after each transformation.

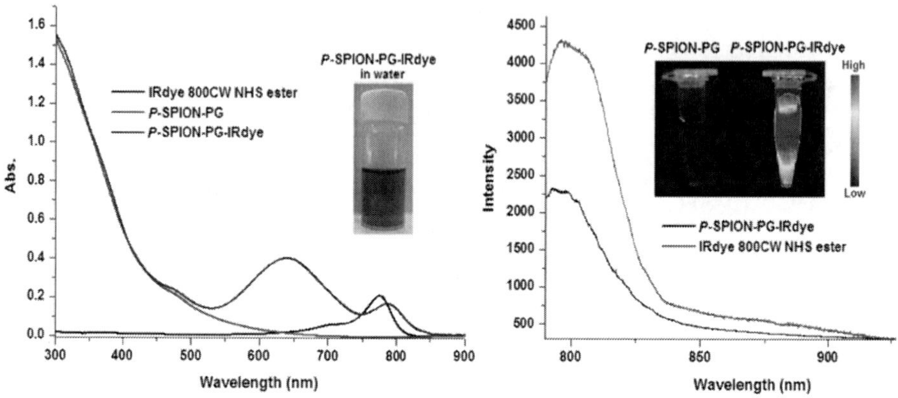

Figure 8. a) UV-Vis-NIR absorption spectra of IRdye 800CW NHS ester, *P*-SPION-PG and *P*-SPION-PG-IRdye in water. Inset: photograph of SPION-PG-IRdye dispersed in water, and b) photoluminescence of *P*-SPION-PG-IRdye in water at the excitation of 785 nm (HORIBA LabRAM). Inset: fluorescence imaging of aqueous solutions of *P*-SPION-PG and *P*-SPION-PG-IRdye (0.5 mg/mL) under ICG mode (Shimadzu Clairvivo OPT; excitation: 785 nm; emission filter: 845 nm; exposure time: 1 sec).

CONCLUSION

We have prepared hydrophilic SPIONs by co-precipitation and polyol methods (*C*- and *P*-SPIONs, respectively), and conducted direct PG grafting on them. The *C*- and *P*-SPIONs, and *C*- and *P*-SPION-PGs were characterized qualitatively by FTIR, STEM, and DLS, and quantitatively by TGA. Since *P*-SPION-PG exhibited high solubility and excellent stability even in a buffer solution, it was separated according to the size by size exclusion chromatography. The separated *P*-SPION-PG showed size-dependant transverse relaxivity (r_2), because the larger magnetic particles are prone to possess higher magnetic moments and hence distort the magnetic field more efficiently to enhance r_2. PG coating provided *P*-SPION with not only excellent solubility and stability in a physiological medium, but also good extensibility to add more functions. In view of active cancer targeting, RGD peptide was immobilized covalently on *P*-SPION-PG through multi-step organic transformations.

The resulting *P*-SPION-PG-RGD demonstrated targeting specificity towards cancer cells overexpressing $\alpha_v\beta_3$-integrin such as A549 and U87MG cancer cells.

In addition, *P*-SPION-PG was conjugated with NIR fluorescent dye through amide linkage. The *P*-SPION-PG-IRdye with strong NIR emission showed great potential for optical/MR bimodal imaging. The PG grafting and further surface functionalization provide a convenient and effective strategy to construct SPION-based platform for a broad range of biomedical imaging and therapy.

EXPERIMENTAL SECTION

Synthesis of *P*-SPION-PG-NH_2

P-SPION-PG-N_3 (8.0 mg) and triphenylphosphine (30 mg, 0.092 mmol) were dissolved in a mixture of DMF (3.0 mL) and water (0.50 mL) by bath sonication, and then concentrated ammonia (28 – 30%, 0.2 mL) was added dropwise. The mixture was stirred at room temperature for 30 h and dialyzed against water to remove DMF. After removal of white precipitates by filtration, the clear brownish filtrate was concentrated by rotary evaporation. It was characterized by FTIR (Figure 7b).

Synthesis of *P*-SPION-PG-IRdye

P-SPION-PG-NH_2 (6.0 mg) was incubated with LI-COR® IRdye 800CW NHS ester (0.2 mg) in pH 8.0 phosphate buffer (0.1 M) at room temperature for 2 h in the dark according to the manufacturer's instruction. To remove unbound dye, the mixture was purified in water by repeated redispersion/centrifugation cycles. It was characterized by FTIR (Figure 7c).

ACKNOWLEDGMENTS

This work was supported financially by Science and Technology Incubation Program in Advanced Region (JST), Industrial Technology Research Grant Program (NEDO), and Grant-in-Aid for Challenging Exploratory Research (JSPS).

REFERENCES

[1] Amstad, E.; Textor, M.; Reimhult, E. *Nanoscale* 2011, *3*, 2819.
[2] Xie, J.; Liu, G.; Eden, H. S.; Al, H.; Chen, X. *Acc. Chem. Res.* 2011, *44*, 883.
[3] Hao, R.; Xing, R.; Xu, Z.; Hou, Y.; Gao, S.; Sun, S. *Adv. Mater.* 2010, *22*, 2729.
[4] Latham, A. H.; Williems, M. E. *Acc. Chem. Res.* 2008, *41*, 411.
[5] Laurent, S.; Forge, D.; Port, M.; Roch, A.; Robic, C.; Elst, L. V.; Muller, R. N. *Chem. Rev.* 2008, *108*, 2064.
[6] Jun, Y.; Lee, J.-H.; Cheon, J. *Angew. Chem. Int. Ed.* 2008, *47*, 5122.
[7] Gupta, A. K.; Gupta, M. *Biomaterials* 2005, *26*, 3995.
[8] Colombo, M.; Carregal-Romero, S.; Casula, M. F.; Gutierrez, L.; Morales, M. P.; Bohm, I. B.; Heverhagen, J. T.; Prosperi, D.; Parak, W. J. *Chem. Soc. Rev.* 2012, *41*, 4306
[9] C. Boyer, M. R. W. V. B. J. L. T. P. D. *NPG Asia Mater.* 2010, *2*, 23.
[10] Amoozgar, Z.; Yeo, Y. *Wiley Interdiscip. Rev. Nanomed. Nanobiotechnol.* 2012, *4*, 219.
[11] Zhao, L.; Takimoto, T.; Ito, M.; Kitagawa, N.; Kimura, T.; Komatsu, N. *Angew. Chem. Int. Ed.* 2011, *50*, 1388.
[12] Takimoto, T.; Chano, T.; Shimizu, S.; Okabe, H.; Ito, M.; Morita, M.; Kimura, T.; Inubushi, T.; Komatsu, N. *Chem. Mater.* 2010, *22*, 3462.
[13] Miguel-Sancho, N.; Bomati-Miguel, O.; Colom, G.; Salvador, J.-P.; Marco, M.-P.; Santamaria, J. *Chem. Mater.* 2011, *23*, 2795.
[14] Zhao, L.; Chano, T.; Morikawa, S.; Saito, Y.; Shiino, A.; Shimizu, S.; Maeda, T.; Irie, T.; Aonuma, S.; Okabe, H.; Kimura, T.; Inubushi, T.; Komatsu, N. *Adv. Funct. Mater.* 2012, *22*, 5107.
[15] Maurizi, L.; Bisht, H.; Bouyer, F.; Millot, N. *Langmuir* 2009, *25*, 8857.
[16] Cai, W.; Wan, J. *J. Colloid Interface Sci.* 2007, *305*, 366.
[17] Wan, J.; Cai, W.; Meng, X.; Liu, E. *Chem. Commun.* 2007, 5004.
[18] Zhao, L.; Takimoto, T.; Kimura, T.; Komatsu, N. *J. Indian Chem. Soc.* 2011, *88*, 1787.
[19] Na, H. B.; Palui, G.; Rosenberg, J. T.; Ji, X.; Grant, S. C.; Mattoussi, H. *ACS Nano* 2012, *6*, 389.
[20] Huang, J.; Bu, L.; Xie, J.; Chen, K.; Cheng, Z.; Li, X.; Chen, X. *ACS Nano* 2010, *4*, 7151.
[21] Jun, Y.; Huh, Y.-M.; Choi, J.; Lee, J.-H.; Song, H.-T.; Kim, S.; Yoon, S.; Kim, K.-S.; Shin, J.-S.; Suh, J.-S.; Cheon, J. *J. Am. Chem, Soc.* 2005, *127*, 5732.

[22] Sisson, A. L.; Steinhilber, D.; Rossow, T.; Welker, P.; Licha, K.; Haag, R. *Angew. Chem. Int. Ed.* 2009, *48*, 7540.

[23] Yeh, P.-Y. J.; Kainthan, R. K.; Zou, Y.; Chiao, M.; Kizhakkedathu, J. N. *Langmuir* 2008, *24*, 4907.

[24] Wei, H.; Insin, N.; Lee, J. Y.; Han, H.-S.; Cordero, J. M.; Liu, W.; Bawendi, M. G. *Nano Lett.* 2012, *12*, 22.

[25] Zheng, W.-W.; Hsieh, Y.-H.; Chiu, Y.-C.; Cai, S.-J.; Cheng, C.-L.; Chen, C. *J. Mater. Chem.* 2009, *19*, 8432.

[26] White, M. A.; Johnson, J. A.; Koberstein, J. T.; Turro, N. J. *J. Am. Chem. Soc.* 2006, *128*, 11356.

[27] Meinhardt, T.; Lang, D.; Dill, H.; Krueger, A. *Adv. Funct. Mater.* 2011, *21*, 494.

[28] Chen, K.; Xie, J.; Xu, H.; Behera, D.; Michalski, M. H.; Biswal, S.; Wang, A.; Chen, X. *Biomaterials* 2009, *30*, 6912.

[29] Decuzzi, P.; Ferrari, M. *Biomaterials* 2007, *28*, 2915.

[30] Xiong, J.-P.; Stehle, T.; Zhang, R.; Joachimiak, A.; Frech, M.; Goodman, S. L.; Arnaout, M. A. *Science* 2002, *296*, 151.

[31] Cai, W.; Chen, X. *Nat. Protoc.* 2008, *3*, 89.

[32] Meng, S.; Su, B.; Li, W.; Ding, Y.; Tang, L.; Zhou, W.; Song, Y.; Caicun, Z. *Med. Oncol.* 2011, *28*, 1180.

[33] Xu, Q.; Liu, Y.; Su, S.; Li, W.; Chen, C.; Wu, Y. *Biomaterials* 2012, *33*, 1627.

[34] Giordano, R. J.; Cardo-Vila, M.; Lahdenranta, J.; Pasqualini, R.; Arap, W. *Nat. Med.* 2001, *7*, 1249.

[35] Kim, J.; Piao, Y.; Hyeon, T. *Chem. Soc. Rev.* 2009, *38*, 372.

In: Magnetic Nanoparticles
Editor: Nora P. Sabbas
ISBN: 978-1-63117-434-6

Chapter 5

HIGH MAGNETIZATION SUPERPARAMAGNETIC CORE-SHELL NANOPARTICLES

Mohamed Abbas[1,2,3]***, Md. Nazrul Islam***[1]
and CheolGi Kim[1,2*]

[1]Department of Materials Science and Engineering,
Chungnam National University, Yuseong-gu, Daejeon, Korea
[2]Department of Emerging Material Science,
Daegu Gyeongbuk Institute of Science and Technology (DGIST),
Daegu, Korea
[3]CeramicsDepartment, National Research Centre, Cairo, Egypt

ABSTRACT

Magnetic nanoparticles and nanocomposites have been special focus of high research interest due to their high magnetic moment and bio-affinity surface properties and promising applications in nanoscience, nanotechnology and bio-applications. These properties can be obtained by synthesizing various magnetic nanoparticles and core-shell type magnetic nanocomposites. In this research, we synthesized high magnetization superparamagnetic nanoparticles and core-shell

* Corresponding author; email ID: cgkim@cnu.ac.kr.

nanocomposites by chemical and sonochemical method for using as carriers/labels in bio-sensing purpose.

Iron oxide magnetic nanoparticles (NPs) have been synthesized by sonochemical method using inexpensive and non-toxic metal salts as reactants. Transmission electron microscopy (TEM) data demonstrated that the particles were narrow range in size distribution with 11 nm average particle size and spherical in shape. The magnetization curve from vibrating sample magnetometer (VSM) measurement shows that as-synthesized NPs were nearly superparamagnetic in magnetic properties with very low coercivity, and magnetization values were 80 emu/g.

Monodisperse magnetite nanocubes with uniform particle size of about 80 nm have been synthesized in aqueous medium by sonochemical method. The magnetic characterization of the NPs reveals saturation magnetization of 91 emu/g at 5 K for as-synthesized sample and 94.8 emu/g for the sample which annealed at the temperature of 600 °C in a vacuum chamber. However, the saturation magnetization has been observed to decrease with further increase in annealing temperature and this has been attributed to the presence of a thin magnetic dead layer at the surface caused by shape anisotropy distortion and broken exchange bonds, and spin canting on the surface of the particles in addition to formation of a small amount of maghemite phase.

Highly crystalline and monodisperse cobalt ferrite ($CoFe_2O_4$) nanoparticles have been synthesized via rapid one-pot sonochemical techniques and without subsequent calcination. The size of $CoFe_2O_4$ nanoparticle was controlled in the range from 20 to 110 nm based on the solvent medium used in the synthesis process. Furthermore, the evolution from spherical to cubic morphology of cobalt ferrite is achieved by simply changing the solvent medium from aqueous to alcoholic medium. High saturation magnetization (Ms) and high coercivity (Hc) values of 87 emu/g and 1610 Oe, respectively were obtained for the $CoFe_2O_4$ NPs.

For synthesis of core-shell type of silica coated iron oxide magnetic (Fe_3O_4@SiO_2 core-shell) NPs, sono-chemical approach was applied using inexpensive and non-toxic chemicals. TEM data demonstrated that the thickness of silica coating on iron oxide magnetic NPs 10-15 nm in average. The magnetization curve from VSM measurement shows that the magnetization has also been decreased of as synthesized silica coated iron oxide NPs compared to freshly prepared bare iron oxide magnetic NPs, which is also a evidence of synthesizing of Fe_3O_4@SiO_2 core-shell NPs.

Besides the synthesis of metal oxide core-shell NPs, high magnetization FeCo nanoparticles with different Fe/Co ratios have been successfully synthesized by surfactant free simple modified polyol method. In this process, polyethylene glycol was used as a solvent media and it has been found to play a key role to act as a reducing agent as well as a stabilizer simultaneously. TEM data suggest that the annealed FeCo nanoparticles are of 50–90 nm in size. The physical Property Measurement System (PPMS) reveals that the $Fe_{60}Co_{40}$ composition among all the samples exhibit highest saturation magnetization of 230.14 emu/g at 5K.

In another study, high magnetic monodisperse NiFe NPs with different compositions have been successfully synthesized polyol method. TEM images displayed formation of a thin oxide layer around the nanoparticles, and confirmed by detection of some oxygen element using EDS measurement. The magnetic properties of the synthesized NiFe NPs samples were measured VSM at room temperature, and the saturation magnetization value was found to be iron content dependent.

INTRODUCTION

During the past decade, growing development in the synthesis of nanoparticles obtain much interest due to its tremendous prospects in various sides. This growing interest is not only for its elementary scientific values but also environmental remediation [1–5] and biomedical applications, such as magnetic resonance imaging (MRI) [6–10], cell and protein separation [11–14], and drug delivery [15], as well as ever-increasing technological applications such as their tunable magnetic properties [16–18] and their wide use in catalysis [19], magnetic recording [20], high performance electromagnetic [21] and spintronic devices [22]. It is essential to have large amounts of magnetic nanoparticles stably dispersed in water for these diverse ecological and biomedical applications. Iron Oxide nanoparticles can be synthesized by thermal decomposition of iron acetylacetonate in 2-pyrrolidone, dispersible in water; these materials could be poorly crystalline and/or polydisperse [16, 19]. Recently, high temperature organic solution phase methods developed for the synthesis of monodisperse semiconductor quantum dots have been used to improve material quality for iron oxide systems with encouraging results.

A magnetite (Fe_3O_4) nanocrystal is considered to be the most promising magnetic oxide because of its various technological applications, like data storage [23], sensors [24], spintronics [25-27], drug delivery [28], and environmental protection [29-30]. As the electrical, optical, and magnetic properties of these nanoparticles strongly depend on their dimensions [31], it is required to synthesize uniform size of magnetite with specific physical features like small size, monodisperse, high magnetization, high crystalline with different shapes as well as with chemical features of non-toxicity and hydrophilicity, depending on their application fields.

The monodispersed magnetite nanocubes and nanorods are interesting not only in introducing novel electric and magnetic properties that are different from those observed in their spherical counterparts, but also in serving as building blocks to prepare self-assembled arrays for the applications in the electronics, data storage and biological sensors [32]. Especially, from the biomedical application point of view, the nanocube surface with specified crystal plane is expected to enhance the amount of the immobilization of biofunctional groups, because the surface to whole volume (SV) ratio is higher than other shapes like octahedron, dodecahedron, icosahedron and sphere shape [33].

Many groups have worked to synthesize magnetite nanocubes by solvothermal [34-35] and thermal decomposition methods [36-37]. Even though these methods provide great benefits in the control of size, component and dispersion, but they need much amount of surfactant, long synthesis time, deoxygenated protection and process in organic phase. However, the physicochemical effects of sonochemical method causes the formation, growth and implosive collapsing of bubbles in liquid medium [38], and thus brings uniform size distribution, high surface area, fast reaction time, improved phase purity [39] and hydrophilicity [40].

Recently, two groups have synthesized magnetite nanoparticles in aqueous solution by sonochemical method using the same precursor of iron ($FeSO_4$. $7H_2O$). Mizukoshi and co-workers have synthesized magnetite nanoparticles by using $FeSO_4$. $7H_2O$ in aqueous solution containing NaOH and (PEG-MS), but they have obtained relatively irregular shapes of nanoparticles with small saturation magnetization of 32.8 emu/g at 300 K [41]. Also, Cheng and co-workers have synthesized magnetite nanoplates using $FeSO_4$. $7H_2O$ in aqueous solution containing NaOH and NaAc. They have succeeded in producing a new shape of nanoplate magnetite, but their sizes are not uniform and saturation magnetization is relatively small, 54 emu/g [42].

Cobalt ferrite ($CoFe_2O_4$) nanoparticle is considered as one of the most promising magnetic oxide materials, because of its high coercivity (Hc), moderate saturation magnetization (Ms), large magnetostrictive coefficient, remarkable mechanical hardness and high chemical stability [43]. All of these properties make it as a potential material for different electronic applications, such as recording media and noncontact torque sensors [44-45]. Furthermore, recent studies reported on the utilization of $CoFe_2O_4$ nanoparticles in different applications, such as catalyst, hyperthermia treatment, magnetic resonance imaging, and biosensor [46-48]. Most of these applications require particles with highly crystalline structure, uniform size and shape distribution, because the electrical, optical and magnetic properties depend strongly on the dimensions of the nanoparticles [49].

Many groups have worked to synthesize cobalt ferrite nanoparticles using different methods such as, hydrothermal, polyol, sol-gel, and coprecipitation [43, 46, 50, 51]. Though all these methods provide good benefits in synthesis of $CoFe_2O_4$ NPs with different structures, but in most cases the $CoFe_2O_4$ NPs obtained were severely aggregated with nonuniform shape and size distributions. Furthermore, the nanoparticles in most cases need a subsequent annealing process to be well crystalline. Moreover, some of these methods need the usage of large amounts of surfactant, require deoxygenated protection, and take long synthesis times. However, the sonochemical method is considered as one of the most promising techniques for synthesis of NPs, where the sonochemistry arises from acoustic cavitation phenomenon, that is, the formation, growth, and collapse of bubbles in liquid medium [52]. Furthermore, the extremely high temperature of about 5000 K, high pressure (~20 MPa), and very high cooling rate (~1010 K/s) are supposed to come from the collapse of the bubbles, and thus can obtain extreme reaction conditions which lead to many unique properties of the synthesized particles in sonochemistry [53]. Moreover, the advantages of the sonochemical approach over conventional methods in the synthesis of metal oxide NPs, including more uniform size distributions, a higher surface area, faster reaction time, and improved phase purity, have been recognized by many research groups [54].

During the past decade, growing development in the synthesis of core-shell nanoparticles (NPs) and nanocomposites obtain much interest due to its tremendous prospects in various sides of modern science and technology. This growing interest is for its biomedical applications, such as magnetic resonance imaging, cell and protein separation, and drug delivery, environmental remediation. It is essential to have large amounts of magnetic NPs and nanocomposites stably dispersed in water for these diverse ecological and biomedical applications. Recently, several core-shell nanomaterials and nanocomposites has been developed for biosensor application such as silica coating of nanoparticles by the sonogel process [55], sonochemical method [56], microemulsion route [57-58].

Magnetic nanoparticles with high saturation magnetization are considered to be very important for magnetic energy, data storage applications [59-62], and biomedical applications such as magnetic separation and drug delivery [63, 64]. Especially, higher saturation magnetization of nanoparticles is required for easy manipulation for targeted delivery of drugs in specific sites through magnetic field assistance [65]. One of the most feasible materials is body centered cubic (bcc) FeCo magnetic nanoparticles because of its high saturation magnetization (Ms= 240 emu/g), Curie temperature and high permeability properties [66, 67].

Several groups have synthesized FeCo nanoparticles using different techniques such as, thermal decomposition [68-71], wet chemical processes (by means of a reducing agent such as borohydride) [72, 73], and chemical vapor deposition [74]. Among the various methods, the polyol process, which involves reduction of a metal salt with a diol, typically ethylene glycol, diethylene glycol, or a mixture of both [75], is believed to be one of the most appropriate methods, because of the possibility to control the experimental conditions kinetically and to scale up easily [76]. Also polyol process is preferred over the other techniques as this does not emit toxic gases like carbon monoxide as in the case of thermal decomposition or contaminate the product as in the case of borohydride reduction [77, 78].

Recently, the Ni-Fe nanoparticles have received great attention because of its diverse practical applications in the fields such as magnetic materials, catalyst, and medicin application [79]. Especially for bioapplications, the magnetic nanoparticles should be highly crystalline, monodisperse, have high magnetization value, and superparamagnetic properties so as to make them suitable for applications such as targeted drug delivery, hyperthermia and magnetic resonance imaging enhancement [80]. Furthermore, the nickel iron nanoparticles containing 65-90 wt% nickel is well known as permalloy, and is important because of its magnetic permeability and low hysteresis loss [81].

Many groups have synthesized NiFe nanoparticles using different methods such as hydrothermal method, hydrogen reduction of Fe and Ni inorganic salts, reverse micelle techniques, and sonochemical decomposition [81-86]. Even though all these methods provide good benefits in synthesis of NiFe NPs with different structure and properties, but the obtaining of NiFe nanoparticles with highly crystalline structure, monodisperse, uniform shape and size distribution with good magnetic properties by easy process and without using of surfactant, and complicated procedures is still of high interest. However, the polyol method which involves reduction of metal salts with a diol, typically ethylene glycol, diethylene glycol, or a mixture of both is believed to be one of the most appropriate methods for synthesis of nanoparticles. Where, by using the polyol process we can dispense for using of surfactant, because the polyethylene glycol (PEG) plays a triple role as high-boiling solvent, reducing agent, and stabilizer to efficiently control the particle growth and prevent interparticle aggregation due to steric interactions, in addition to its hydrophilic properties [87].

DISCUSSION

1. Nanospheres and Nanocubes Cores

1.1. Iron Oxide (Fe_3O_4) Nanospheres

In this study, we report a new facile synthetic route for the preparation of superparamagnetic monodisperse Fe_3O_4 nanospheres by sonochemical method using inexpensive and non-toxic metal salts as reactants. By optimizing the preparation conditions such as concentration of Ammonia and reaction time, we have successfully synthesized monodisperse Fe_3O_4 NPs sphere with the average size of 11 nm, which implies that the modified sonochemical method could be a good candidate to prevent toxicity problems in typical sonochemical synthesis process. Superparamagnetic nature of the nanoparticles was observed by measuring magnetic hysteresis loop using vibrating sample magnetometer (VSM). It is revealed from X-ray diffraction (XRD) and transmission electron microscopy (TEM) results that the as synthesized nanoparticles have small size distribution and moderate monodispersity in both polar and non polar solvents. Furthermore, the single crystalline behavior of the as prepared nanoparticles was confirmed from TEM observation.

1.1.1. Experimental Procedure

The NPs synthesis was carried out under an ambient temperature. Analytical reagent grade chemicals were used without further purification. Prepared mixture of 2.81 g $FeCl_3.6H_2O$ (99%, Aldrich) and 1.72 g $FeCl_2.4H_2O$ (90%, Aldrich) in 100 ml distilled H_2O was sonicated for 30 minutes. The ultrasonic processor (Vibra Cell-VCF 1500, Sonics and Materials) with a maximum power of 1500 watts generating capacity was used in this experiment. The sonoreactor was equipped with a titanium horn having 5 cm^2 of irradiating surface area and piezoelectric transducer supplied by a 20 kHz generator. Then mixture was sonicated in ultrasonic processor for 75 minutes. 15 ml of ammonia was injected in the reaction vessel after 15 minutes of starting of ultrasonication. Finally, the obtained mixture was washed with distilled water and ethanol for several times and dried in a vacuum evaporator.

The crystal structure of the synthesized NPs was analyzed using an XRD system (Rigaku, D/MAX-RC, Cu Kα radiation and a Ni filter). The magnetic properties were characterized by means of a Lake Shore 7400 series vibrating sample magnetometer (VSM) at room temperature. The size of the synthesized NPs was characterized using a typical transmission electron microscopy (TEM, Technai F-20, operated at 200 kV). The size distributions of the particles were measured from enlarged TEM images.

1.1.2. Characterization

The XRD patterns of the sample (Figure 1.A) shows that the peaks can be indexed to (220), (311), (400), (422), (511), (440) and (533) planes of a cubic unit cell of Fe_3O_4 based on the comparison with the standard pattern of Fe_3O_4 (JCPDS card no. 79-0418). The results of EDS analysis are shown in figure 1 (B). When the EDS intensity factors are taken into accounts, the NPs contained 78.5% by atomic ratio of iron (Fe) and 21.5% by atomic ratio of oxygen (O) which can be assumed that the obtained nanoparticles were Fe_3O_4 NPs.

To investigate the magnetic properties of the NPs, room temperature magnetization measurements have been carried out by using a VSM in an external magnetic field range between -13 kOe and +13 kOe. Figure 2 shows the M-H hysteresis loops for the sample. The hysteresis loops reveal that the as synthesized NPs have very low Hc, 10 Oe, which is near to superparamagnetic behavior. The magnetization value at the maximum applied field was 80 emu/g. Magnetic susceptibility of as synthesized NPs has also been calculated. From the calculation, it is revealed that the as synthesized NPs have high susceptibility. The estimated mass susceptibility is $Xg = 5.71 \times 10\text{-}4\ m^3/kg$.

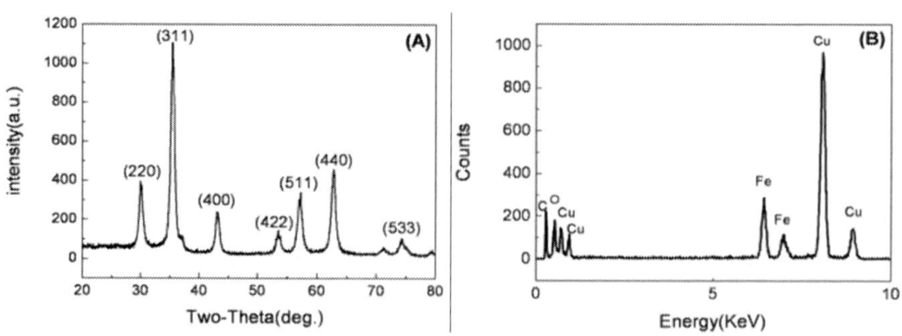

Figure 1. (A) XRD patterns of the sample shows the evident peaks for magnetite phase and (B) EDS patterns of Fe_3O_4 NPs which clearly shows the presence of Fe and O elements in the NPs. Cu and C peaks are due to carbon copper grid [88].

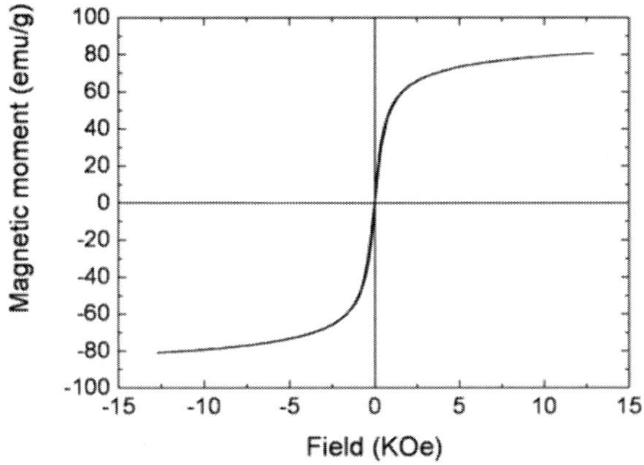

Figure 2. The magnetization curves of the Fe_3O_4 NPs measured at room temperature [88].

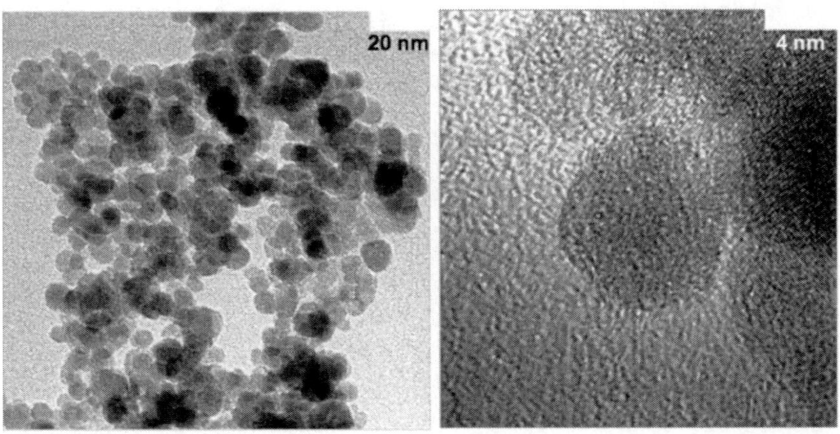

Figure 3. TEM images of NPs (a) Fe_3O_4 NPs (b) Single Fe_3O_4 NP is spherical in shape [88].

Figure 3 (a-b) shows the TEM images of Fe_3O_4 NPs. The average particles size of the sample was estimated to be about 11 nm. It is also revealed that as synthesized iron oxide NPs were round in shape. The size and morphology of the particles were analyzed from TEM images and summarized in figure 5 as the size distribution diagram. Based on the size distribution information, it is clear that the size of as prepared Fe_3O_4 NPs has the relatively narrow range in particles size distribution.

1.2. Magnetite (Fe_3O_4) Nanocubes

Herein we report a one-pot new facile synthetic route for the preparation of high moment Fe_3O_4 nanocube by sonochemical method using $FeSO_4$. $7H_2O$ in aqueous solution containing only NaOH, and without deoxygenated conditions. The crystalline structure and shapes of synthesized nanocubes were examined by XRD, SEM, TEM, and the saturation magnetization was analysed as functions of measuring and annealing temperatures in terms of crystalline phases.

1.2.1. Sonochemical Formation of Fe_3O_4

The chemical reaction driven by intense ultrasonic waves, which are strong enough to produce cavitation, causes oxidation, reduction, dissolution, decomposition and hydrolysis. Ultrasonic irradiation of aqueous liquids generates free radicals of $H^{\cdot}$ and $OH^{\cdot}$ Radicals [92]. These radicals can recombine to return their original form or combine to produce H_2 and H_2O_2, and these resultant strong oxidants and reductants in turn are utilized during various sonochemical reactions in aqueous solutions [39, 89].

$$H_2O \rightarrow H^{\bullet} + OH^{\bullet} \tag{1}$$

$$H^{\bullet} + H^{\bullet} \rightarrow H_2 \tag{2}$$

$$OH^{\bullet} + OH^{\bullet} \rightarrow H_2O_2 \tag{3}$$

$$FeSO_4\ (s) \rightarrow Fe^{+2}{}_{(aq)} + (SO_4)^{-2}{}_{(aq)} \tag{4}$$

Then the oxidant H_2O_2 thus generated is suggested to initiate the oxidation of Fe^{+2}

$$2\ Fe^{+2} + H_2O_2 \rightarrow 2\ Fe^{+3} + 2OH^{-} \tag{5}$$

$$Fe^{+2} + 2Fe^{+3} + 8\ OH^{-} \rightarrow Fe_3O_4 + 4H_2O \tag{6}$$

1.2.2. Experimental

1.2.2.1. Materials

Iron (II) sulphate heptahydrate ($FeSO_4.7H_2O$) (98.0 ~ 102.0 %), sodium hydroxide (NaOH) (bead 98.0 %), and ethyl alcohol (C_2H_5OH) (99 %) were purchased from Samchun pure chemical. Co, Ltd. All chemicals were of analytical reagent grade and used as received without any further purification and the synthesis process was carried out under an ambient temperature.

1.2.2.2. Synthesis of Magnetite Nanocubes

In a typical synthesis, 2.31 gm of $FeSO_4$. $7H_2O$ was dissolved in 90 mL distilled water for 10 minutes using magnetic stirrer, and then sonicated using an ultrasonic processor for 75 minutes. 9 mL of 3M NaOH was injected in the reaction after 15 minutes of starting ultrasonication. A schematic diagram of the sonochemical experiment setup is shown in Figure 1. The ultrasonic processor (Vibra Cell-VCF 1500, Sonics and Materials) with a maximum power of 1500W was used in this experiment. The sonoreactor was equipped with a titanium horn having 5 cm^2 of irradiating surface area, and piezoelectric transducer supplied by a 20 kHz generator was immersed below the surface of the sonicated liquid. Finally, the obtained mixture was washed and sonicated (using cleaner SH-3400) for five times in water and ethanol while collecting the precipitate using a magnet, and then dried in a vacuum evaporator. The as-synthesized iron oxide nanoparticles were annealed in the temperature range of 400~ 700 °C in a vacuum chamber for 2 hours.

1.2.3. Characterization

The crystal structure of the as-synthesized and annealed nanoparticles was analyzed by X-ray powder diffraction (XRD. Regaku D/max-250) with Cu K α radiation (λ=1.540562 Å) and in the 2θ range from 20 to 80 degrees. The operational voltage and current were kept at 40 kV and 300 mA, respectively. The size and morphology of the nanoparticles were characterized using Scanning electron microscopy (SEM, JEOL JSM-7000F) and Transmission electron microscopy (TEM, The Tecnai G2 F20 operated at 200 kV). The chemical composition was analyzed by the EDS detector embedded on the TEM. The magnetic properties of the nanoparticles were measured by means of a Physical Property Measurement System (PPMS, Quantum Design Inc) at different temperatures in the range of 5 ~ 300 K.

1.2.4. Results and Discussion

1.2.4.1. Structure Characterization

The X-ray diffraction pattern of as-synthesized magnetite sample is shown in Figure 4. The peaks can be indexed at 30.1°, 35.4°, 37.0°, 43.0°, 53.39°, 56.9°, 62.6°, 70.9°, 73.9° and 78.9°, corresponding to the crystal planes of (220), (311), (222), (400), (422), (511), (440), (620), (533) and (444), respectively. The sharper and stronger peaks indicate the formation of iron oxide with a cubic inverse spinel structure, which are consistent with the standard data for magnetite (JCPDS card no.00-019-0629). Figure 4(b-d) shows the annealing temperature effects on the peaks, where the intensity of the peak slightly increases with annealing temperature until 600 °C, then decreases at 700 °C. Besides, new peaks at (203), (116), (216), (109) and (226) are also noticed, corresponding with the maghemite phase of γ-Fe_2O_3 (JCPDS card no. 01-076-3169).

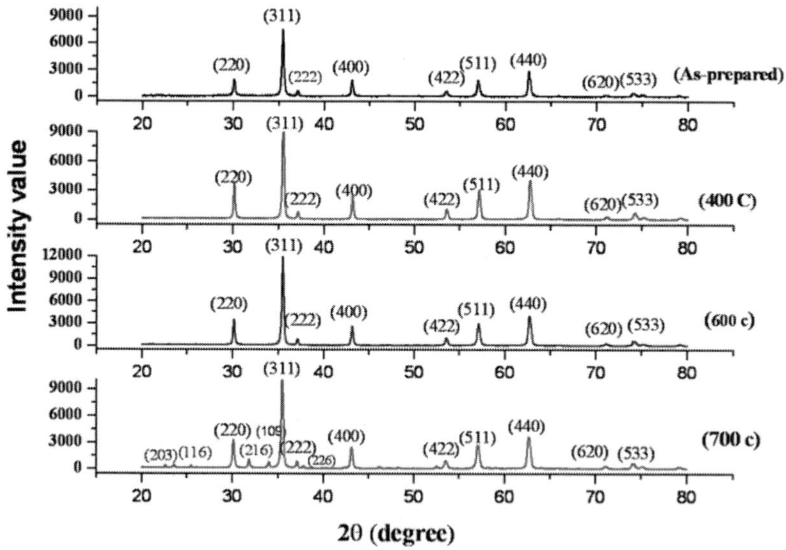

Figure 4. XRD patterns of magnetite nanocubes; (a) as-synthesized, (b) annealed at 400 °C, (c) annealed at 600 °C and (d) annealed at 700 °C [89].

Using Rigaku, Jade 9 program we determined the percentage of maghemite phase as 1.7 %. The presence of this small amount of maghemite, even though we used a vacuum chamber in the annealing of the magnetite sample, may be attributed to the so called topotactic transformation. In such case, the possibility for oxidation is the diffusion of iron rather than the presence of additional oxygen, and the previous studies have shown that iron diffusion is enough to explain the kinetics of the oxidation of magnetite to maghemite [93].

EDS analysis in Figure 3a shows that the synthesized nanoparticles are iron oxide. Here we can get the Fe_3O_4 formula of the nanoparticles, containing 75.3 % and 24.7 % for iron and oxygen (estimated as weight %) concentrations, where the atomic weight of iron and oxygen is 55.8 and 15.9, respectively. So it is noted that the as-synthesized nanoparticles are single phase magnetite (Fe_3O_4). Also, the annealed samples at 400 °C, 600 °C and 700 °C show nearly the same results of iron and oxygen concentration.

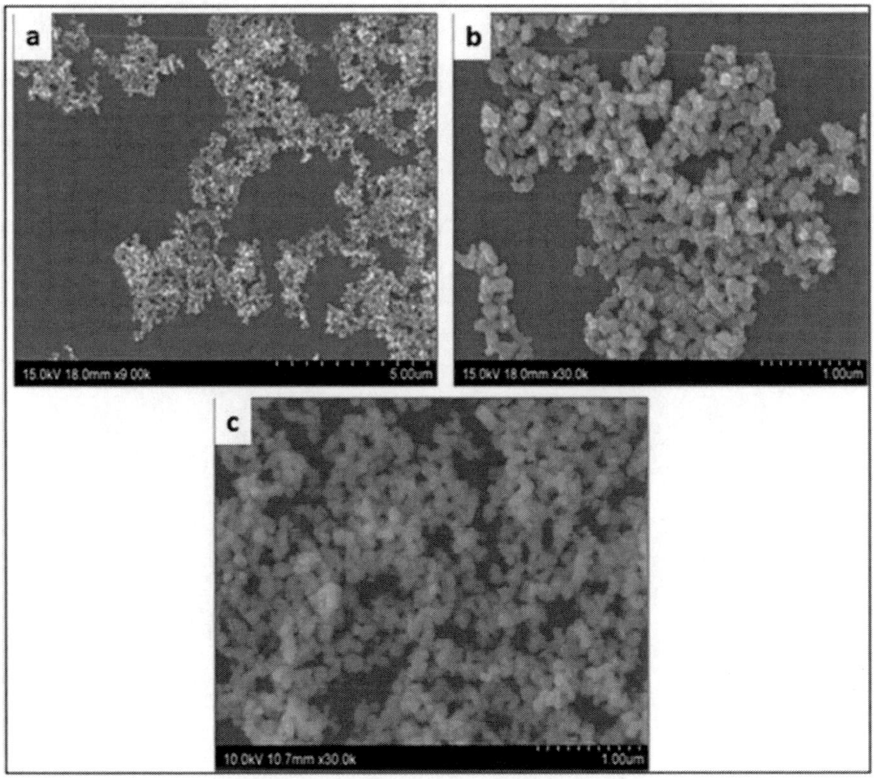

Figure 5. SEM images of magnetite nanocubes at different magnifications; (a) as-synthesized NPs with low magnification, (b) as-synthesized NPs with high-magnification. (c) annealed sample[89].

1.2.4.2. Morphology Characterization

The shape of the magnetite nanoparticles was shown in Figures (4, 5, 6 and 7) observed by SEM and TEM for the as-prepared and annealed magnetite nanoparticles. The SEM images in Figures 4(a, b) reveal that these particles have a relatively narrow size distribution and nearly uniform cube shape. Even after the annealing at 700 °C, the SEM image of the sample shows only a little agglomeration of the particles, but still there is no distinctive growth in size as shown in Figure 5c.

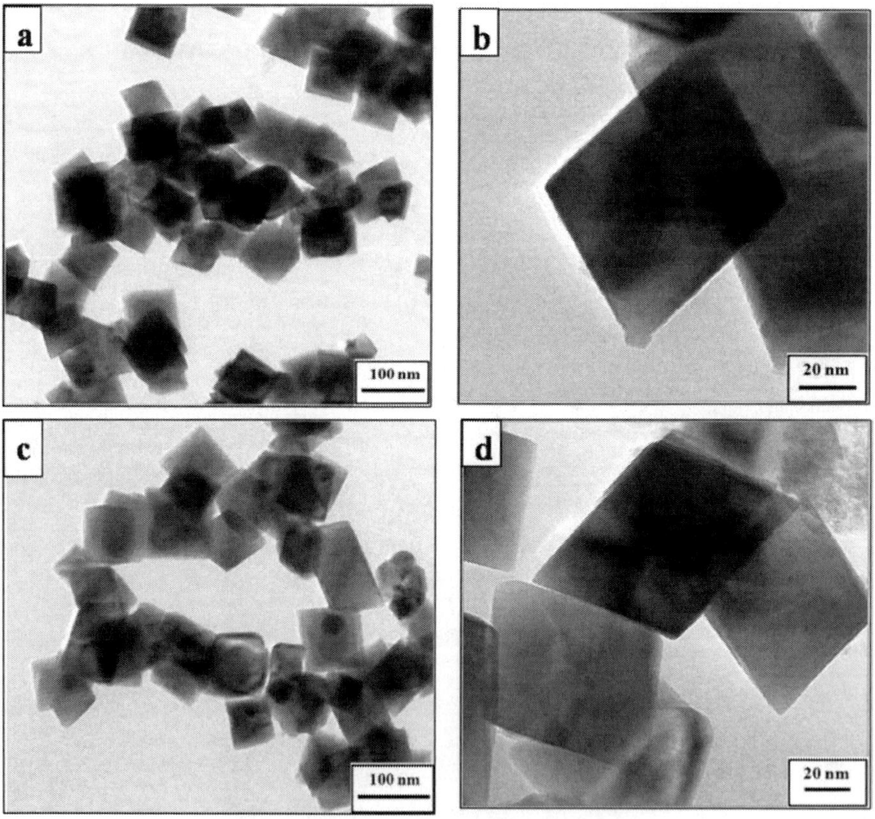

Figure 6. TEM images of the annealed magnetite nanocubes at (a, b) 400 °C, and (c, d) 600 °C[89].

Figure 6 shows the low-magnification TEM image of the as-prepared magnetite nanoparticles, where the cubic shape with well crystallinity is revealed. This is consistent with the sharp and stronger peaks obtained from the XRD data and thus confirms the cubic inverse spinel structure of the magnetite. Furthermore, the electron diffraction pattern, in the inset of Figure 6a, exhibits a good crystalline nature of magnetite, corresponding to a cubic structure. The high-resolution TEM (HRTEM) image of the nanoparticles, as shown in Figure 6b, reveals the high crystallinity of the nanoparticles with clear-cut edges and average particle size of about 80 nm.

Figures 6 and 7 show the TEM images of the annealed magnetite nanocubes at 400 °C, 600 °C and 700 °C, respectively. Figure 6(a, b) shows the low- and high magnification TEM images of the annealed magnetite at 400 °C which reveals a small increment in size to be nearly in the range of 80 to 85 nm with high crystallinity structure. Whereas, Figure 6(c, d) shows the TEM images of the annealed sample at 600 °C with size ranging from 85 to 90 nm and are marked by very high crystallinity structure with sharp- cut edges.

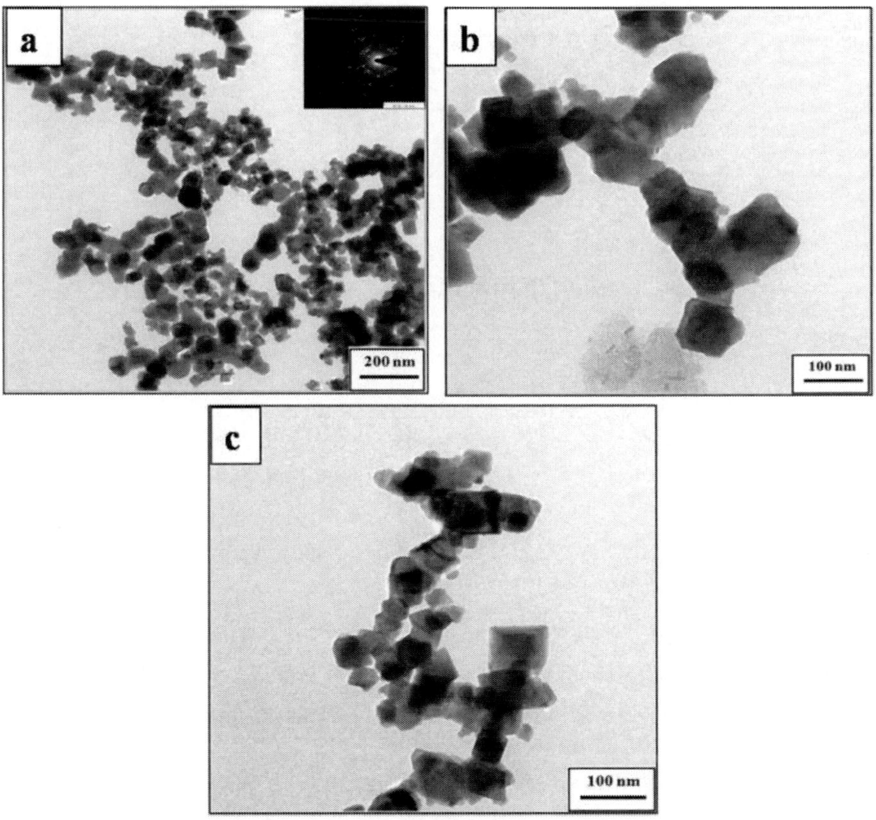

Figure 7. TEM images of the annealed magnetite nanocubes at 700 °C with different magnification (a, b) Low-magnification and (c) high- magnification. Inset in (a) represents electron diffraction pattern [89].

On the other hand, increasing the annealing temperature up to 700 °C resulted in a slight distortion in the shape of the magnetite nanocube, as shown in Figure 7a. This distortion is clearly visible in the magnified TEM images in Figure 7(b, c) where in the surfaces of the particles look like melted or broken. Moreover, the size of the nanoparticles in case of annealed 700 °C sample is in the range of 40 to 90 nm with a broad size distribution comparing with the uniform size and shape in the other annealed samples of 400 °C, 600 °C and the as-synthesized magnetite. Furthermore, the electron diffraction pattern, in the inset of Figure 7a, also exhibits the distortion or the low crystallinity structure of the magnetite nanocube. The decrease in particle size and the change in particle shape which resulted from the effect of annealing at 700 °C are agreed with the previous report [94].

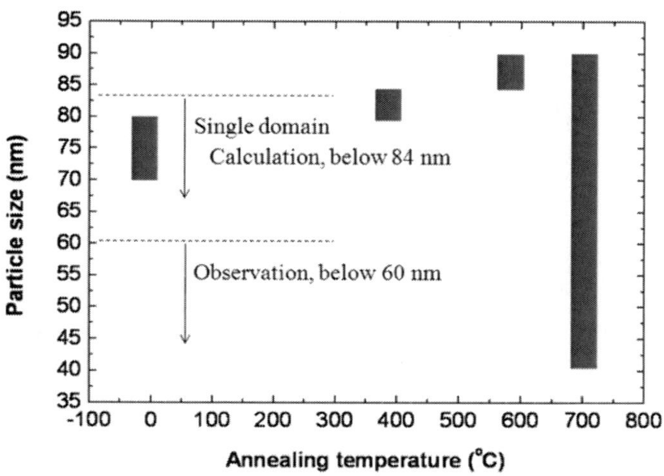

Figure 8. The diagram for particle size distribution of as-synthesized and annealed magnetite nanocubes obtained from TEM images [89].

Figure 8 shows the diagram for particle size distribution of the as-synthesized and annealed magnetite nanocubes deduced from the analysis of the TEM images, and it explains the behaviour of nearly constant increment of size with annealing temperature up to 600 °C as illustrated above. Nevertheless, though there are little variations in the size and the size-distributions for the samples of as-synthesized and annealed magnetite nanocubes, it may by and large be concluded from the above that the sonochemical technique in the present study produced magnetite nanoparticles in cube shape only.

Ultrasonics irradiation in the sonochemical technique provides rather unusual reaction conditions such as extremely high temperatures (about 5000 K) and high pressures (about 20 MPa) in liquids combined with a rapid cooling rate (about 10^{10} K/S) for a short duration, which can otherwise not be realized by other methods [53]. Also, it is believed that the ultrasonics irradiation significantly enhances the hydrolysis rate, and shock waves can induce unusual morphological changes in the synthesis of metal oxides, leading to more uniform size distribution, a higher surface area, faster reaction time, and improved phase purity [39], in addition to the hydrophilic properties [40]. Thus, if one takes into account the advantages offered by the sonochemical technique and also the higher surface-to-volume ratios shown in Figure4 for evaluation, it can be understood that the most probable shape resulting out of sonochemical technique could be that of the cube only. In accordance with the above, the magnetite nanocubes in the present study clearly satisfy all the above typical characteristics of sonochemical approach particularly to have the uniform size distribution and large surface to volume ratio as evident from the obtained SEM and TEM images, and also the hydrophilicity of our sample which resulted as a reason for using only aqueous medium in the reaction. Further, apart from the fast reaction time (75 minutes for the as-synthesized sample) used in our experiment, the high phase purity and crystallinty of structure were also confirmed from the data of X-ray diffraction. However, in spite of our observations of cube shaped magnetite nanoparticles, based on the existing literature it is still far from reaching any specific conclusion that the sonochemical technique can produce only the cube shape or sphere shape nanoparticles, because different reports in the literature using this technique produce metal or oxide nanoparticles of different shapes like, cube, sphere, plates, etc. [41-42, 53].

1.2.4.3. Magnetic Properties

The magnetic properties of the magnetite nanocube were measured at different temperatures using a VSM in an external magnetic field ranging from −13 kOe to +13 kOe. It is well known that the high crystallinity of the particles leads to formation of magnetite nanocubes with practically no surface spin canting of magnetic moments, and thus justifies the obtained high saturation magnetization values for the as-synthesized particles in the present study [108].

Figure 9 represents the effect of annealing temperature on magnetic properties of magnetite nanocube. The magnetization values of the Fe_3O_4 nanocubes measured at room temperature for the annealed samples at 400, 600 and 700 °C were found to be 88.2, 90.2 and 77 emu/g, respectively. The results indicate that the saturated magnetization value is little increased with increasing annealing temperature up to 600 °C, and then sharply decreased at 700 °C.

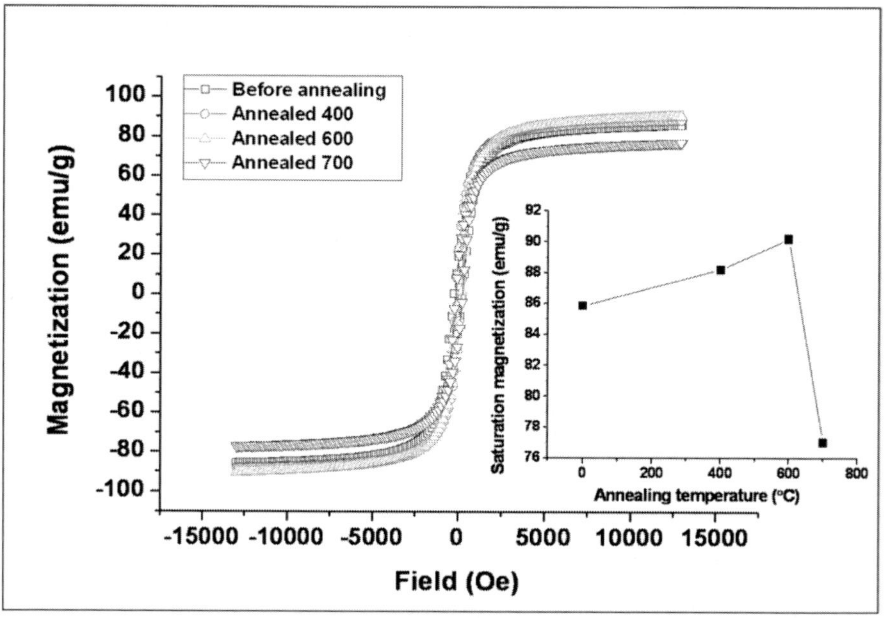

Figure 9. Magnetization curves for magnetite nanocube annealed at different temperature 400, 600 and 700 °C, and measured at 300 K. Inset is the saturation magnetization value of magnetite nanocube at different annealing temperature[89].

As shown in Figures 2, the intensity of the highest peak (311) increases only up to 600 °C annealing temperature. As it is well known that the magnetization value is very much sensitive to the size of the nanoparticles, a little increment in the crystallites size after annealing leads to a corresponding increase in the magnetization value, and thus the observed increase in magnetization values for the annealed samples up to 600 ^{o}C is in accordance with the above consideration and is also well supported by the measured TEM data. It is clearly evident from the TEM images that the size of the magnetite nanocubes increased from nearly 80 nm in case of as-prepared sample to 85 nm in case of 400 °C annealed sample and to nearly 90 nm for the annealed sample of 600 °C. Moreover, the sample which was annealed at 600 °C apparently showed the highest degree of crystallinity and it also showed the highest magnetization value of 94.8 emu/g at 5 K.

On the other hand, the decreased magnetization value at 700 °C annealing temperature may be attributed to different factors. Firstly, as evident from the TEM images in Figure 7 and the diagram of particle size distribution in Figure 8, the sample annealed at 700 °C has a broad size distribution in the range of 40 to 90 nm comparing with the uniform particle size of the other annealed samples at 400 °C and 600 °C or even with the as-synthesized sample. So, the decreasing particle size of the sample annealed at 700 °C was considered to be in a single domain state, as per the previous reports on single domain limit for experimentally obtained cubic magnetite particles below 60 nm [95] and theoretically estimated particle sizes below 84 nm [96], where in the surface spins experience canting and may also lead to formation of a thin magnetic dead layer on the surface [97]. Moreover, decreasing particle size leads to an increasing surface to volume ratio [98] which in turn causes more surface spin disorder and consequently a significant reduction in the magnetization value obtained [99].

The second reason for the decreasing magnetization value in case of sample annealed at 700°C may be concerned with the distortion in the shape of the particle at the surface and its reduced crystallinity, as it is very clear from the TEM images of Figure 7c comparing with high crystallinity and the uniform cubic shapes of the other annealed or as-synthesized samples as in Figures 5, 6. It is well known that the magnetic anisotropy of the particles is dependent on the degree of crystallinity of the nanoparticles and thus a distortion in the shape would naturally cause a reduction in the magnetic moment of the particles [100] and the third reason for the decrease of magnetization could be that of the formation of a small maghemite phase of γ-Fe_2O_3, as revealed in the XRD pattern in Figure 4d, since the magnetization of bulk maghemite is 74 emu/g [101].

As expected, the coercivity increases from 173 Oe to 450 Oe with decrease in the measurement temperature from room temperature (300 K) to 5 K. The reason for such an increase has been reported and may be understood from the Kneller's law [102]. On the other hand, the coercivity values of the annealed magnetite nanocubes are found to be slightly increased from 173 Oe for the as-synthesized magnetite nanocube to 181 Oe for the sample which annealed at 600 °C and then slightly decrease to be 178 Oe with further annealing temperature at 700 °C . The increase and decrease in the coercivity value may be related to the size of the particle. Furthermore, the increase in the coercivity value may be attributed to the increase in the size of the magnetite nanocubes with increasing annealing temperature.

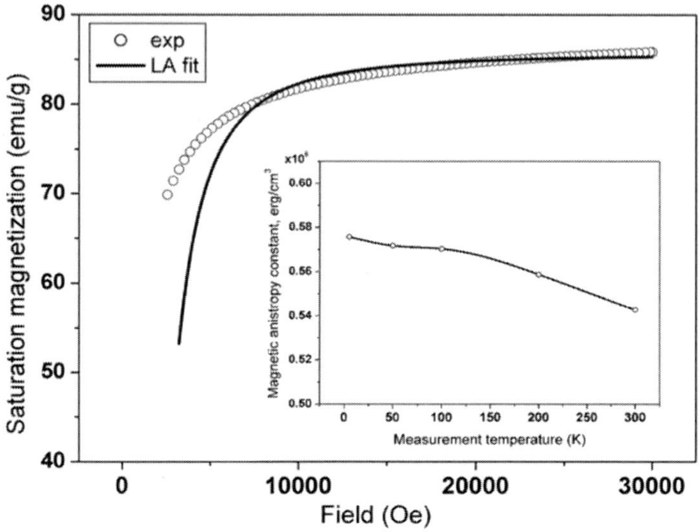

Figure 10. Fit to the Law of approach to saturation (LA) for the as-synthesized magnetite nanocube. Inset is the relation between the magnetic anisotropy constant (K1) at different measurement temperature 5, 50, 100, 200 and 300 K [89].

Typical fitting curve of LA to saturation along with an inset showing the variation of magnetic anisotropy as a function of measurement temperature for the as-synthesized magnetite nanocube is shown Figure 10. The estimated magnetic anisotropy was found to be in the range from 0.57×10^6 erg/cm^3 to 0.53×10^6 erg/cm^3 for 5 K to 300 K, respectively. And thus it is considered as independent of the measurement temperature. This behavior of magnetic anisotropy is attributed to the large surface areas and negligible surface spin canting of our magnetite nanocubes.

1.3. $CoFe_2O_4$ Nanospheres and Nanocubes

In our previous work, we developed a sonochemical method to synthesize highly crystalline magnetite (Fe_3O_4) NPs with uniform shape and size distribution in only aqueous medium. Herein, we successfully synthesized highly crystalline, high magnetization monodisperse $CoFe_2O_4$ NPs with uniform sphere shape in one-step surfactantless sonochemical process without any deoxygenated protection. Furthermore, the synthesis time used in our experiment is only 70 min, and that too is worked out without the need for any subsequent annealing. In addition, since the solvent plays an important role in the reaction not only to control the nucleation and crystal growth but also in the formation of ferrite, the effect of different solvents of aqueous, alcohol, mixed of water/ethanol, and aqueous/PVP is therefore examined on the morphology and the magnetic properties of the cobalt ferrite NPs by various structural and magnetic characterization techniques.

1.3.1. Experimental

1.3.1.1. Materials

Iron (II) sulphate heptahydrate ($FeSO_4.7H_2O$), Cobalt sulphate heptahydrate ($CoSO_4$.7H2O), polyvinyl pyrolidone (PVP), Sodium hydroxide (NaOH), and ethyl alcohol (C_2H_5OH) (99 %) were purchased from Samchun Pure Chemical Co, Ltd. All the obtained chemicals were of analytical reagent grade and were used as received without any further purification, and the synthesis process was carried out under an ambient temperature.

1.3.1.2. Synthesis of Monodisperse Cobalt Ferrite ($CoFe_2O_4$) Nanoparticles

We synthesized cobalt ferrite nanoparticles ($CoFe_2O_4$ NPs) exactly in the same manner as described in our previous work [89] but with a small modification by using different solvent media. In a typical synthesis, 8 mM $FeSO_4$.7H2O and 4 mM $CoSO_4.7H_2O$ was dissolved in suitable amount of distilled water (aqueous medium) for 10 minutes using magnetic stirrer, and then sonicated using an ultrasonic processor for 70 minutes. 3M NaOH was injected in the reaction after 15 minutes of starting ultrasonication. The ultrasonic processor (Vibra Cell-VCF 1500, Sonics and Materials) with a maximum power of 1500W was used in this experiment. The sonoreactor was equipped with a titanium horn having 5 cm2 of irradiating surface area, and a piezoelectric transducer supplied by a 20 kHz generator immersed below the surface of the sonicated liquid. Finally, the obtained mixture was washed and sonicated (using cleaner SH-3400) for five times in water and ethanol while collecting the precipitate using a magnet. It was subsequently dried in a vacuum oven to obtain $CoFe_2O_4$ NPs (herein after referred to as S1). The same procedure was adopted twice again to obtain $CoFe_2O_4$ NPs by changing the solvent medium to ethanol (in this case, the sample is referred to as S2), and to a mixed solution of water/ethanol in 1:1 volume ratio (here, it is referred to as S3) in place of distilled water. Further, in order to understand the role of a stabilizing agent in the synthesis of $CoFe_2O_4$ NPs, we also added 1 gm of polyvinyl pyrolidone (PVP) to the aqueous medium as an additive, and this sample is herein after referred to as S4.

1.3.2. Characterization

The crystal structures of the synthesized $CoFe_2O_4$ NPs were analyzed by X-ray powder diffraction technique (XRD, Rigaku D/max-2500 at a voltage of 40 kV, a current of 300 mA and a scanning rate of 2 °/min with a step size of 0.01°). The size and morphology of the nanoparticles were characterized using transmission electron microscopy (TEM, The Tecnai G2 F20 operated at 200 kV). The chemical composition was analyzed by the energy dispersive X-ray spectrometer (EDS) embedded on the TEM. Fourier transform infrared (FTIR) spectroscopic data was taken in the range from 4000-400 cm^{-1}. The magnetic properties of the synthesized nanoparticles were measured by vibrating sample magnetometer (VSM, Lake Share 7400) with an external magnetic field ranging from –15 kOe to +15 kOe.

1.3.3. Results and Discussion

1.3.3.1. Structure Characterization

The X-ray diffraction patterns of cobalt ferrite NPs synthesized in aqueous medium (S1), alcohol (S2) and mixed solution of water/ethanol in 1:1 ratio (S3) are shown in Figure 11.

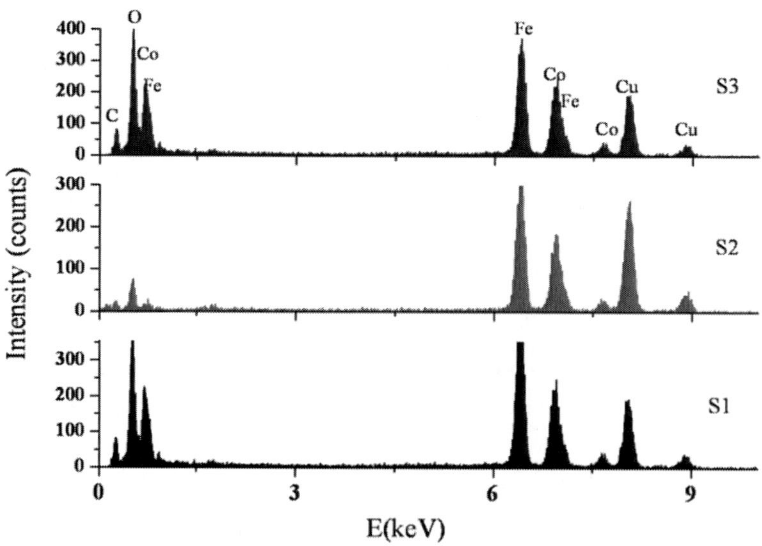

Figure 11. XRD patterns of cobalt ferrite NPs synthesized by sonochemical method in (a) aqueous medium (S1), (b) alcohol medium (S2), and (c) Water: Ethanol (1:1) (S3) [90].

It is found that all the peaks in the three patterns could be indexed to a cubic inverse spinel structure of $CoFe_2O_4$ NPs, which are consistent with the standard data for ferrite phase (JCPDS card no.00-019-0629).

Furthermore, no peaks were detected in the patterns for any impurities which indicate the synthesis process produces high purity cobalt ferrite nanoparticles in a single reaction. Besides, the strong and sharp intensity peaks in case of (S1, S3) samples compared to the peak intensities of (S2) sample imply that the use of either the aqueous medium alone or a mix of water/ethanol medium produces higher crystallinity of the $CoFe_2O_4$ NPs than that of alcohol medium only. In addition, since the crystallite size is inversely proportional to the full width at half maximum (FWHM), the sharper peaks and corresponding smaller peak broadenings of the sample (S3) compared to the FWHMs of (S1 and S2) resulted in larger crystallite size for that sample.

1.3.3.2. Morphology Characterization

The work presented here is based on the modified ultrasonics sonochemistry technique which was previously reported by our group for the synthesis of uniform magnetite nanocube [89]. In this method, the chemical reaction was driven by intense ultrasonic waves which are strong enough to produce cavitation while causing oxidation, reduction, dissolution, decomposition and hydrolysis. Furthermore, the free radicals of H. and OH., which could be generated from the ultrasonic irradiation of aqueous liquids [92], are combined to produce H_2 and H_2O_2 [39], and these resultant strong oxidants and reductants in turn are utilized for the synthesis of NPs [54]. Thus, the proposed mechanism for formation of $CoFe_2O_4$ NPs will be as follows: At first, the two precursors of metals ($FeSO_4.7H_2O$ and $CoSO_4.7H_2O$) we used in our experiment for iron and cobalt, respectively, decompose as a result of these unusual extreme conditions to form the divalent cations of Fe^{+2} and Co^{+2} (Eq. 1-4). Then, the H_2O_2, which was generated as a consequence of combining the two radicals of H. and OH., is suggested to initiate the oxidation of both the divalent cations of Fe^{+2} and Co^{+2} to form Fe^{+3} (Eq. 2) and cobalt oxide (Eq. 5), respectively. And finally, both these products recombine under the influence of ultrasound to form the final structure of cobalt ferrite ($CoFe_2O_4$) nanoparticles (Eq. 6) [90].

$$FeSO_4\,(s) \rightarrow Fe^{2+}(aq) + (SO4)^{2-}(aq) \qquad (1)$$

$$2\,Fe^{2+} + H_2O_2 \rightarrow 2\,Fe^{3+} + 2OH^- \qquad (2)$$

$$Fe^{2+} + 2Fe^{3+} + 8\,OH^{-} \rightarrow Fe_3O_4 + 4H_2O \quad (3)$$

$$CoSO_4\,(s) \rightarrow Co^{2+}(aq) + (SO_4)^{2-}(aq) \quad (4)$$

$$Co^{2+} + H_2O_2 \rightarrow CoO + H_2O \quad (5)$$

$$Fe_3O_4 + CoO \rightarrow CoFe_2O_4 \quad (6)$$

The shapes and sizes of the as-prepared cobalt ferrite nanoparticles synthesized by sonochemical reaction with different solvents used in our experiment are shown in Figures (2, 3, and 4) observed by transmission electron microscopy (TEM). It is apparent that different shapes and sizes of $CoFe_2O_4$ NPs are formed according to the solvent used. The Figure 12 shows the well resolved TEM images of $CoFe_2O_4$ NPs synthesized in aqueous medium with nearly uniform sizes in spherical shape, and the nanoparticles also appear monodisperse with good crystalline structure which agree well with the XRD results discussed earlier. The average size of the particles estimated from the images is about 100 nm with a narrow size distribution.

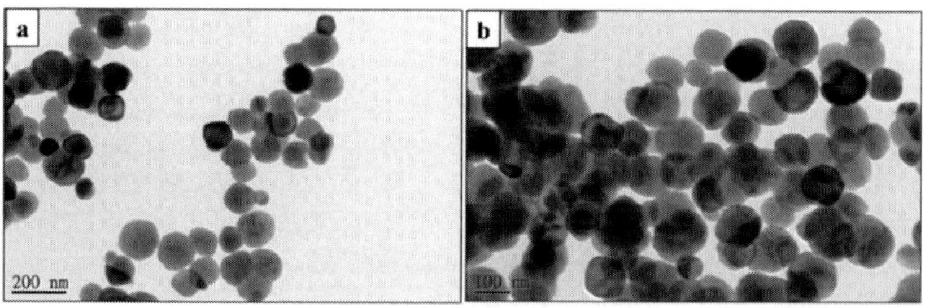

Figure 12. TEM images of cobalt ferrite ($CoFe_2O_4$) NPs synthesized in aqueous medium (S1) [90].

When ethanol is used as solvent, a dramatic decrease in the particle size is obtained to be in the range of 20 nm with evolution of shape from spherical to cubic (Figure 13a, b and c). The reason for such a difference in shape and size of the particles obtained, when we used water and ethanol medium separately as a solvent medium in our reaction, could be due to the high reducing properties of ethanol with respect to water, where the ethanol molecules were probably adsorbed onto the crystal surfaces and thus limit the growth rate, thus leading to a decrease in the particle size and a change in particle shape. Similar results of decreasing the particle size while using ethanol as solvent medium in sonochemical reaction have been reported in the study of magnetite (Fe_3O_4) nanoparticles [53].

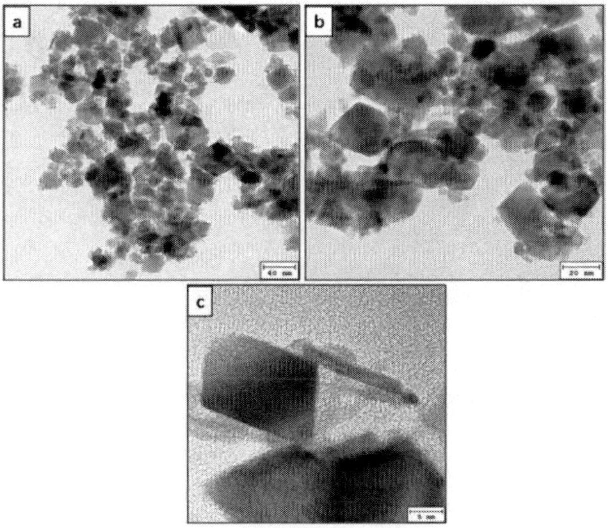

Figure 13. TEM images of cobalt ferrite ($CoFe_2O_4$) NPs synthesized in alcoholic medium (S2) [90].

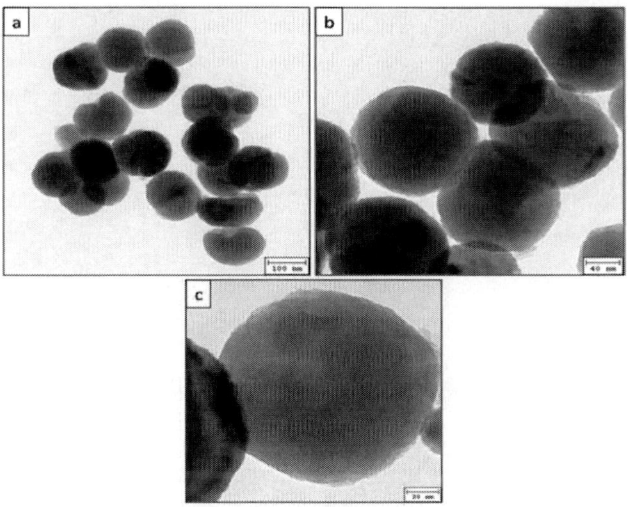

Figure 14. TEM images of cobalt ferrite ($CoFe_2O_4$) NPs synthesized in a mixed solution of Water/Ethanol in 1:1 volume ratio (S3) [90].

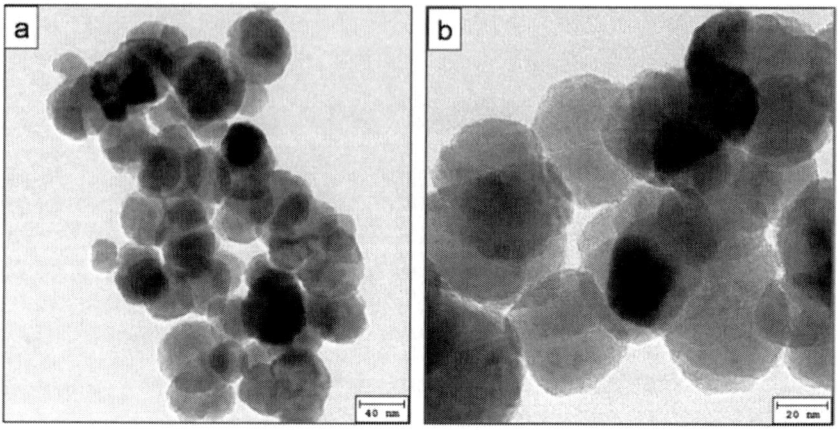

Figure 15. TEM images of cobalt ferrite (CoFe2O4) NPs synthesized by adding 1 gm of PVP to the aqueous solution [90].

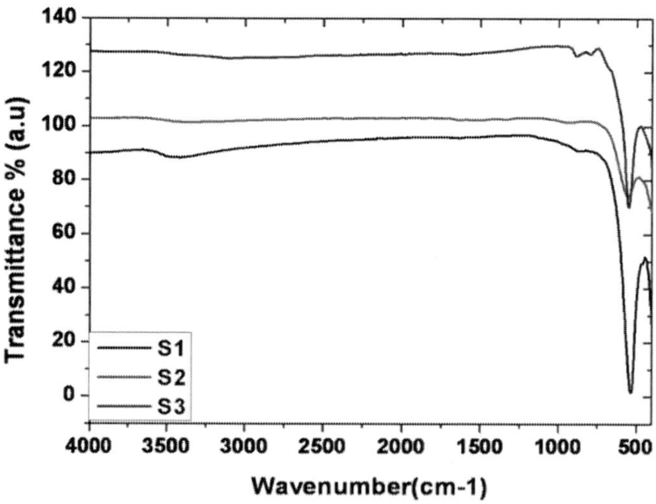

Figure 16. FTIR spectra of cobalt ferrite NPs synthesized by sonochemical method in aqueous medium (S1), alcohol medium (S2), and Water: Ethanol (1:1) (S3) [90].

On the other hand, when we used a mixed solution of water and ethanol as a reaction solvent, a very highly crystalline and monodisperse structure of cobalt ferrite nanoparticles were resulted (Figure 14). Furthermore, the particles obtained in this case have a uniform sphere shape distribution with an average particle size of 110 nm. The improvements in crystalline structure and monodispersity of the synthesized $CoFe_2O_4$ NPs in case of using mixture of water and ethanol solution as solvent media may be attributed to an increased rate of oxidation of Fe^{2+} and consequently an acceleration rate of formation of nanoparticles. Enomoto and coworkers reported the same results in the sonochemical synthesis of ferrite NPs with an increased reaction rate when they used a mixture of water and ethanol rather than using only aqueous medium [53].

In order to study the effect of capping agent in the synthesis process of cobalt ferrite nanoparticles by the sonochemical technique, we added one gram of polyvinyl pyrolidone (PVP) to the aqueous solution prior to the reduction process with sodium hydroxide as a stabilizing polymer. Choosing the PVP as a capping agent in our study is based on its hydrophilic properties and consequently the possibility for its solubility in the aqueous medium. Figure 15 shows the TEM images of the cobalt ferrite nanoparticles obtained after using the PVP. The average size of the particles was dramatically decreased from 100 nm in case of using only aqueous medium to be in the range of 20-40 nm when PVP was added. Such a decrease in size in this case may perhaps be attributed to the addition of a large quantity of PVP (one gram in our case) to the mixture which greatly diminishes the Ostwald ripening process because the additional PVP present in the solution caps many of the free metallic surface sites in the nanoparticles. Consequently, since there are available only a fewer sites for the Ostwald ripening process, the nanoparticles do not increase in size [103].

The composition of as-synthesized $CoFe_2O_4$ NPs was confirmed by EDS analysis for different samples. There are mainly three elements of Fe, Co and O only presents in the sample with desired ratios and thus indicates that the composite is $CoFe_2O_4$ NPs in the three samples. However, the obtained Cu and C peaks in the spectra are due to carbon copper grid employed in the measurements. Further experimental confirmation is obtained through the FTIR spectra for three different samples (S1, S2 and S3), as shown in Figure 16.

All the three samples contain large peak around 542 cm^{-1} which related to the vibration of Fe-O functional group matching well with the characteristic peak of ferrite [104]. Also, all the curves contain another peak at around 3500-3000 cm^{-1} which could be attributed to the stretching of OH groups.

3.3.3. Magnetic Properties

The magnetic properties of the samples of $CoFe_2O_4$ NPs (S1, S2, S3 and S4) were measured at room temperature using VSM in an external magnetic field ranging from –13kOe to + 13kOe, as shown in Figure 17. It can be seen from the loops that the applied field is not enough to obtain saturation of magnetization for any of the samples. However, saturation magnetizations for all these samples were calculated by using the linear fits of M versus 1/H curves as 77.01, 37.78, 92.57 and 75.81 emu/g for S1, S2, S3 and S4 samples respectively, and the same were shown in the inset of Figure8 for different samples. On the other hand, the maximum magnetizations were estimated from the loops for the four samples (S1, S2, S3 and S4) as 71, 33.8, 87 and 70.5 emu/g, respectively. Indeed, the magnetization values of S1 and S4 are close to the theoretical value of bulk cobalt ferrite ($CoFe_2O_4$) of 73 emu/g and that of S3 sample is even higher than the theoretical bulk value. This higher value may be ascribed to the larger particle size with negligible surface spin disorder. This argument is well supported by the higher degree of the crystallinity of these synthesized $CoFe_2O_4$ NPs, as observed through their corresponding XRD patterns, resulting in negligible surface spin canting and thus justifying the higher values of magnetization [104].

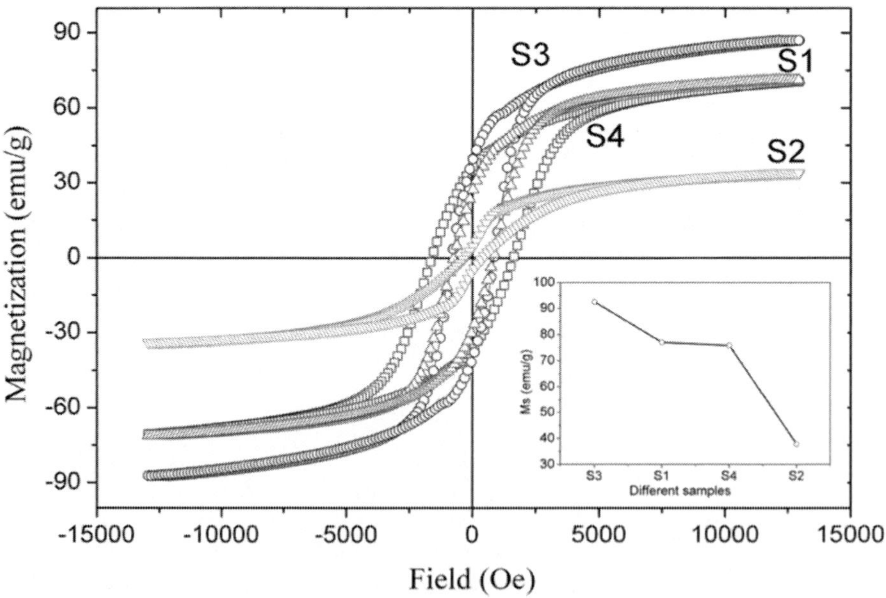

Figure 17. Magnetization curves of cobalt ferrite NPs synthesized in aqueous medium (S1), alcohol medium (S2), mixed solution of Water: Ethanol (1:1) (S3), and PVP added to the aqueous solution (S4). The inset represents the calculated values of saturation magnetization for the four samples using M versus 1/H linear fitting curves [90].

At the same time, the dramatic decrease in the magnetization value of the sample synthesized in alcohol medium (S2) is mainly attributed to its small particle size and poor crystallinity since it is well known that the magnetization value is very much sensitive to the size and crystallinity of the particles. Further, decrease in the particle size leads to an increase in surface to volume ratio [98], which in turn causes more surface spin disorder [99], and consequently a significant reduction in the magnetization value obtained.

Magnified view of the curves near low magnetic field to make the coercivity value more clear for different samples of synthesized $CoFe_2O_4$ NPs, and the inset represents the values of coercivity for the same samples (732, 328, 807 and 1610 Oe for S1, S2, S3 and S4 samples, respectively). The samples which used aqueous medium in full or part exhibit similar values of coercivity at 732 Oe (S1) and 807 Oe (S3), whereas the PVP added sample (S4) displays maximum value of coercivity at 1610 Oe. Interestingly, rather low value of coercivity is obtained for S2 sample at 328 Oe. The reason for such a huge difference in the coercivity value among different samples may be ascribed to surface pinning that arises due to missing coordination of oxygen atoms and the shape effect of the nanaoparticles. Unlike the curved topography in spherical $CoFe_2O_4$ NPs which display larger coercivities, the cubic $CoFe_2O_4$ MNPs are hypothesized to fewer missing oxygen atoms and less surface pinning, resulting in lower coercivity for the cubic structures [105]. Song et al. reported similar results of variations in coercivity with shape structure from sphere to cube [106]. Thus, it may be pointed out that the shape effect plays an important role in determining the coercivity similar to the size effect on the magnetization value of these nanoparticles.

2. Core@Shell Nanoparticles

2.1. Silica Encapsulated Iron Oxide Nanospheres

In this study, we report a facile synthetic route for the preparation of nearly superparamagnetic Fe_3O_4@SiO_2 core-shell Nanospheres by sono-chemical method using inexpensive and non-toxic metal salts as reactants. By optimizing the preparation conditions such as concentration of Ammonia, metal salts and reaction time, we have successfully synthesized Fe_3O_4@SiO_2 core-shell NPs with the average size of 30 nm, which implies that this modified simple sono-chemical method could be a good candidate to synthesize various nanocomposites. Nearly superpara-magnetic nature of Fe_3O_4@SiO_2 core-shell Nanospheres was observed by measuring magnetic hysteresis loop using vibrating sample magnetometer (VSM). It is revealed from X-ray diffraction (XRD) and transmission electron microscopy (TEM) results that the as synthesized $Fe_3O_4@SiO_2$ core-shell NPs have small size distribution and moderate monodispersity in both polar and non polar solvents.

2.1.1. Experimental Procedures

2.1.1.1. First Experiment

Two steps sono-chemical method has been implied for synthesizing $Fe_3O_4@SiO_2$ core-shell NPs. In the first step, Prepared mixture of 3 g $FeCl_3.6H_2O$ (99%, Aldrich) and 2 g $FeCl_2.4H_2O$ (90%, Aldrich) in 120 ml distilled H_2O was sonicated for 60 minutes. 20 ml of aqueous 28-30% ammonia was injected in the reaction vessel after 10 minutes of starting of ultrasonication [88]. The detailed synthesis process has been described in previously published paper. The obtained mixture was washed with distilled water and ethanol for several times and dried in a vacuum evaporator.

2.1.1.2. Second Experiment

Freshly prepared Fe_3O_4 NPs (50 mg) were dispersed in cyclohexane (50 mL) under sonication for 10 min to get the Fe_3O_4 NPs solution (1 mg/mL) in the presence of oleic acid (2 mL). In another 500 mL erlenmeyer flask, 250 mL of cyclohexane was mixed with 12 g of Igepal CO-520 and magnatically stirred vigorously for 5 minutes. Then Fe_3O_4 NPs solution was added into this solution and the resulting mixture was further magnetically stirred for 10 min. Aqueous 28-30% ammonia (3 mL) was added to this solution drop wise via syringe. After magnetically stirring for 5 min, tetraethylorthosilicate (1 mL; TEOS) was added drop wise via a syringe. The mixture was stirred for more than 15 h in the room temperature. Methanol (100 mL) was added to the solution to form dark precipitates. The dark precipitate was washed in ample amount of ethanol. The synthesized $Fe_3O_4@SiO_2$ core-shell NPs has been dried in the vacuum dryer for characterization of sample.

2.1.2. Characterizations

The XRD patterns of the sample are shown in Figure 18. The XRD analysis shows that the peaks represented for silica encapsulated magnetite phase (JCPDS powder diffraction file 27-1402) and the peaks can be indexed to (111) (220), (311), (400), (422), (511), (440) and (533) planes of a cubic unit cell. From XRD measurements it is evident that the as synthesized NPs were silica encapsulated Fe_3O_4.

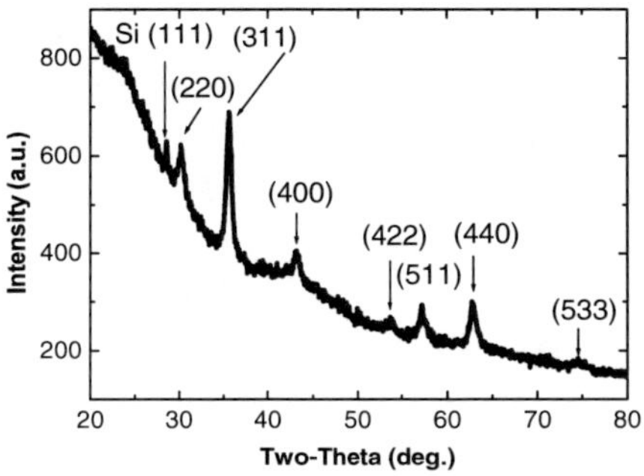

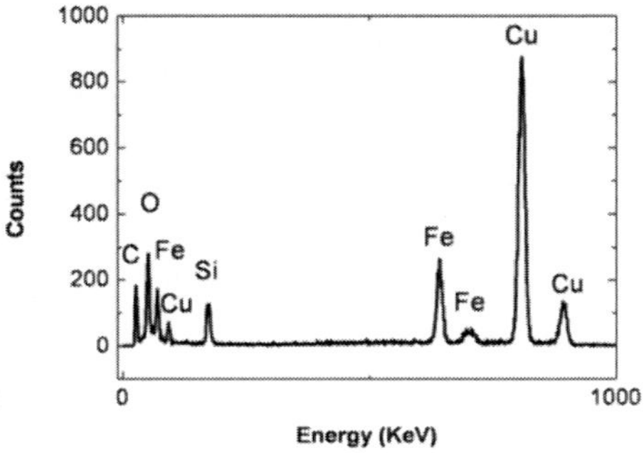

Figure 18. XRD patterns (left) of the sample shows the evident peaks for $Fe_3O_4@SiO_2$ core-shell NPs phase and EDS patterns (right) of $Fe_3O_4@SiO_2$ core-shell NPs which clearly shows the presence of Fe, O and Si elements in the NPs. Cu and C peaks are due to carbon copper grid [107].

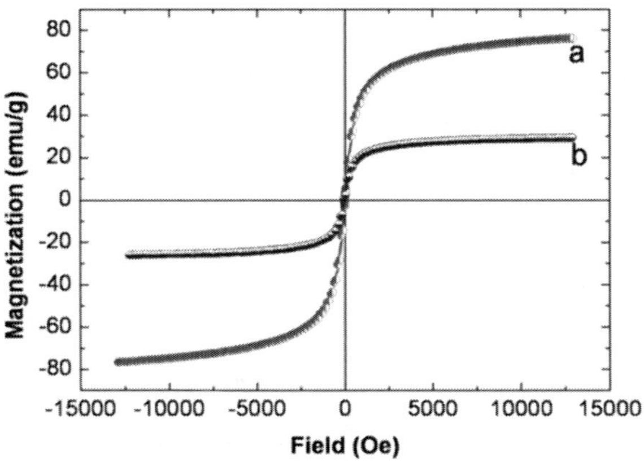

Figure 19. The magnetization curves of the (a) Fe_3O_4 NPs and (b) $Fe_3O_4@SiO_2$ core-shell NPs measured at room temperature [107].

The presence of silica along with iron oxide NPs in the synthesized $Fe_3O_4@SiO_2$ core-shell NPs sample has also been confirmed by the evident peaks of silica, iron and oxide from EDS analysis. When the EDS intensity factors are taken into accounts, the Fe_3O_4 NPs contained 78.5% by atomic ratio of iron and 21.5% by atomic ratio of oxygen which can be assumed that the obtained freshly prepared NPs which were used for silica coating were Fe_3O_4 NPs.

To investigate the magnetic properties of the $Fe_3O_4@SiO_2$ core-shell NPs, room temperature magnetization measurements have been carried out by using a vibrating sample magnetometer (VSM) in an external magnetic field range between -13 kOe and +13 kOe. Figure 3 shows the M-H hysteresis loops for the freshly prepared Fe_3O_4 NPs (Figure 19a) and Fe_3O_4@SiO_2 core-shell NPs (Figure 19b).

The hysteresis loops reveal that the as synthesized Fe_3O_4@SiO_2 core-shell NPs were nearly superparamagnetic. The magnetization value for Fe_3O_4 NPs at the maximum applied field was 77 emu/g, on the other hand, for Fe_3O_4@SiO_2 core-shell NPs was 28 emu/g. The M-H hysteresis loops of VSM measurement in the Figure (19b) shows that the magnetization value has shown significant decrement of Fe_3O_4@SiO_2 core-shell NPs (28 emu/g) compared to bare Fe_3O_4 NPs (77 emu/g), which is also an evidence of synthesizing of silica coated iron oxide NPs. It is thought that a part of silica is bonded on the surface of magnetite particles through the chemical connection (Fe-O-Si). The magnetic moment of iron ions connected with Si on the surface of magnetite particles disappears through the Fe-O-Si connection and the magnetism decreases through the silica coating. Moreover, similar results of decreasing the magnetization value with silica coating have been reported.

Figure 20 shows the TEM images of core-shell Fe_3O_4@SiO_2 NPs. The size and morphology of the nanoparticles were analyzed from TEM images. The average particles size of freshly prepared Fe_3O_4 NPs was estimated to be about 12 nm in diameter. It is revealed that as synthesized iron oxide NPs and core-shell Fe_3O_4@SiO_2 NPs were round in shape. It is also reveals that after encapsulation of Fe_3O_4 NPs with silica, the size of Fe_3O_4@SiO_2 NPs was estimated to be from 25 to 35 nm in diameter with 10-15 nm of silica shell thickness.

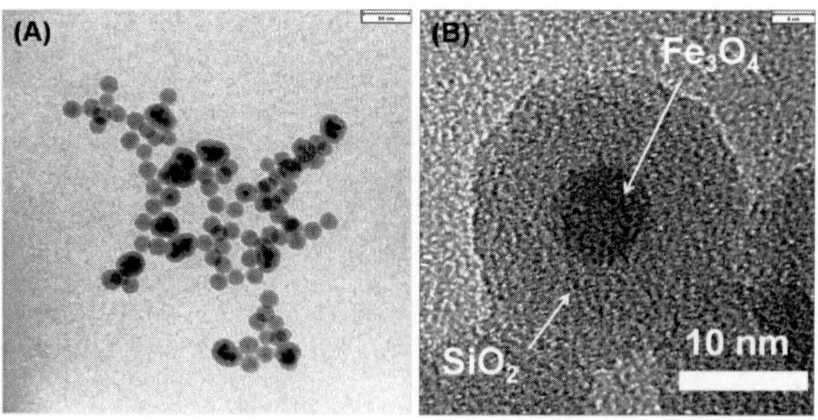

Figure 20. TEM images of NPs (A) Fe_3O_4@SiO_2 core-shell NPs (B) Single Fe_3O_4@SiO_2 core-shell NPs is spherical in shape[107].

2.2. Oxide Coated FeCo Nanoparticles

Recently, many groups have been synthesized FeCo nanoparticles using polyol method [67, 97]. Even though these methods provide high saturation magnetization which reaches in some case to 221 emu/g, but particle sizes are larger than 100 nm after annealing the sample at 600 °C. Herein, by optimizing the Fe/Co ratio, concentration of hydroxyl ion and the reaction condition in a surfactant free facile one- pot modified polyol method, we successfully synthesized small size and air stable FeCo nanoparticles with high saturation magnetization. The crystalline structure and shapes of the synthesized nanoparticles were examined by XRD, TEM, EDS and the saturation magnetization was measured using PPMS.

2.2.1. Experimental

FeCo nanoparticles with different compositions and size prepared by modified polyol method. The details of chemicals used are provided in the supporting materials.

2.2.1.1. Synthesis of FeCo Nanoparticles

FeCo nanoparticles were prepared with different chemical compositions such as $Fe_{20}Co_{80}$, Fe_{50} Co_{50}, $Fe_{60}Co_{40}$, and $Fe_{70}Co_{30}$. Firstly, in round bottom flasks, mixtures of ($FeCl_2.4H_2O$) and ($Co(Ac)_2.4H_2O$) with a suitable amount of Polyethylene glycol (PEG) which was used as dispersing media was prepared in different compositions. For all the compositions, the pH of the solutions was adjusted in between 10-11 by adding NaOH prior to reduction process. Then the PEG-metal salts solution was heated up to 300 °C, and was continuously stirred using a magnetic stirrer, refluxed at this temperature for 2 hours. In all cases the solution turned black within a few minutes of reaching refluxing temperature. Then the solutions cooled down to room temperature and followed by 20 minutes in ultrasonication and washed several times using ethanol and water and then collected by using the magnet.

2.2.2. Results and Discussion

2.2.2.1. Structure Characterization

The X-Ray diffraction patterns of synthesized FeCo samples are shown in Figure 21. It was apparent that three characteristic peaks for FeCo indexed at value of 44.82, 65.2 and 82.66 degrees corresponding to the crystal planes of (110), (200) and (211) were observed for all FeCo of different compositions.

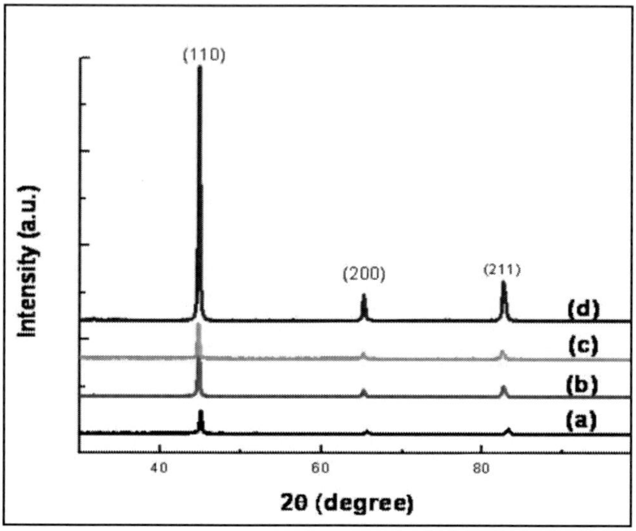

Figure 21. XRD patterns for different compositions of the 600 °C annealed FeCo nanoparticles; (a) $Fe_{20}Co_{80}$, (b) $Fe_{50}Co_{50}$, (c) $Fe_{60}Co_{40}$, and (d) $Fe_{70}Co_{30}$ [87].

The strong and sharp diffraction patterns match only with that of the α-FeCo of body centered cubic structure (bcc) consistent with the standard data for FeCo (JCPDS card no. 00-044-1433), and no additional peaks for iron oxide or cobalt oxide were observed. Typical EDS analysis of the annealed FeCo samples in Figure 2 reveal that the materials are mainly composed of Fe and Co metals only while displaying a small percentage (≈2- 3%) of oxygen, which could not be sufficient enough to be identified through XRD patterns, for all samples with different Fe/Co ratio.

2.2.2.2. Morphology Characterization

Figure 22 (a, b) shows the typical TEM images of the as-prepared FeCo nanoparticles which reveal that the as-synthesized nanoparticles are monodisperse with10 nm average size and spherical of shape. This smaller size of the synthesized nanoparticles has been successfully obtained mainly by controlling the reaction conditions such as reaction time, PEG, Fe/Co ratio, optimization of the concentration of hydroxyl ions, and the ultrasonication procedure. The optimum choice of the concentration of hydroxyl ions is not only important in size control but also in prohibiting the material from formation of additional phases. For example, we found that by decreasing the concentration of hydroxyl ions in solution, a cobalt ferrite phase was formed in the reaction solution. Also, when the reaction time is increased for more than 2 hours, it resulted in increment of particle size to more than 100 nm. This has been demonstrated in Figure 22 (c, d) and also in supporting materials which shows the TEM image of the annealed sample where the particles were still spherical in shape but with increased sizes to about 50-90 nm.

The stability of the sample annealed at 600 °C was investigated by measuring the XRD analysis and the magnetization after 3 months of synthesizing samples. The sample has not shown any visible changes in colour, shape, structure and magnetization when stored at room temperature in air. That means, the samples exhibit a strong surface stability and are not likely get oxidized even for at least 3 months of contacting with air. This resistance to oxidation may be due to formation of a thin passive oxide layer around the particles as was observed from the EDS analysis, on the presence of nearly 2- 3 % of oxygen [65], and was also evident by its appearance in the TEM image of Figure 22e. Further, this may also be due to the formation of another thin layer of PEG on the particle surface, where it is well known that the poly ethylene glycol has strong reducing properties and consequently protects it from oxidation.

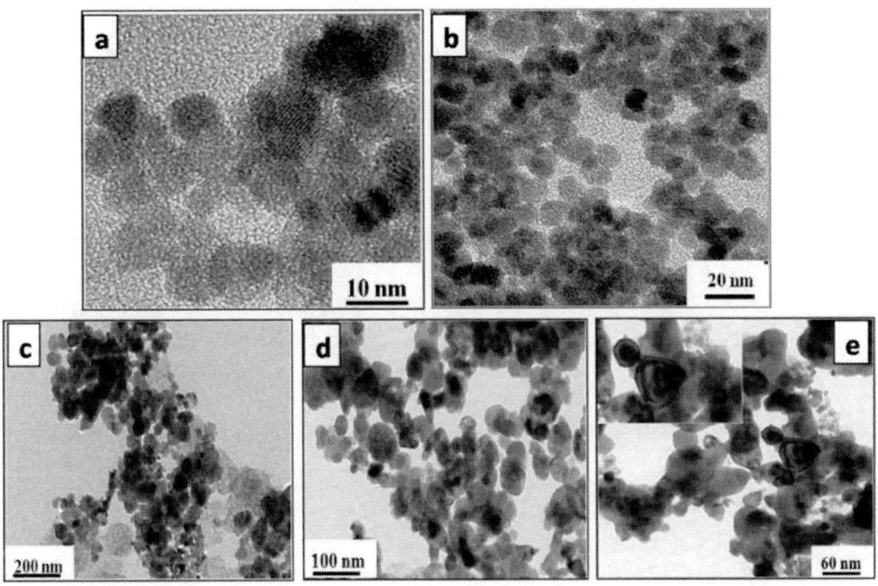

Figure 22. TEM images for FeCo nanoparticles (a) As-synthesized with high magnification, (b) As-synthesized with low magnification, (C) Annealed at 600 °C with low magnification, (d) Annealed at 600 °C with high magnification, and (e) thin oxide layer [87].

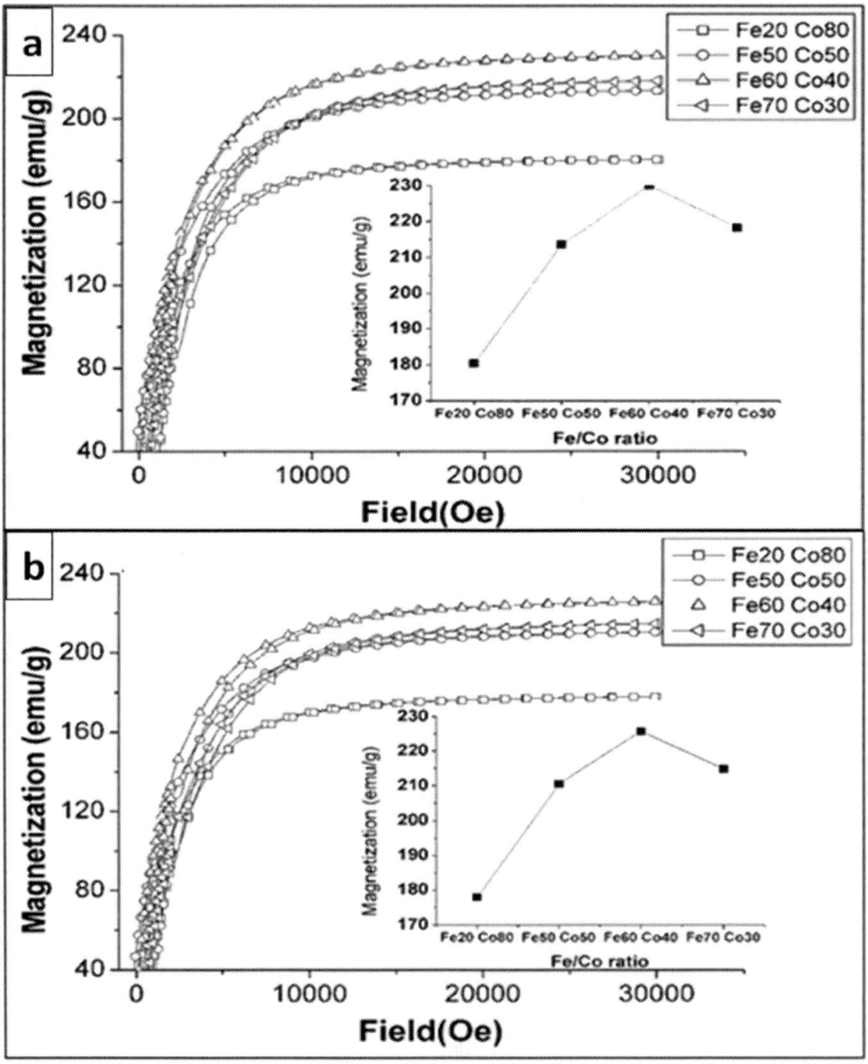

Figure 23. The magnetization curves of FeCo nanoparticles for different compositions annealed at 600 °C; (a) room temperature and (b) 5 K using PPMS [87].

2.2.2.3. Magnetic Properties

The magnetization of FeCo alloy depends on the ratio of Fe and Co atoms. Figure 23 (a, b) shows the magnetization curve of FeCo nanoparticles with different Fe/Co ratio measured at room temperature and 5 K. As expected, the samples measured at lower temperatures resulted in higher values of saturation magnetization. In the inset in each figure, the magnetization trend as a function of Fe/Co ratio has been shown. In both the cases, an increase in the saturation magnetization with iron content was observed until the optimized Fe/Co ratio with $Fe_{60}Co_{40}$ composition annealed at 600 °C and then it decreased. For this sample, we have obtained the saturation magnetization of 230.14 emu/g, which is close to the value of bulk FeCo (~ 240 emu/g) [63].

Also, there is an agreement between our results of the highest saturation magnetization value obtained (in cases of $Fe_{60}Co_{40}$ and $Fe_{70}Co_{30}$) and the reported one for Fe/Co ratios. The lowest saturation magnetization of 180 emu/g is obtained for the sample with high cobalt content ($Fe_{20}Co_{80}$), and this may be obviously due to relatively lesser contribution by low magnetic moment metallic cobalt compared to that of high magnetic moment metallic iron. Besides, the coercivity is observed to increase from 170 Oe for the sample with low cobalt content ($Fe_{70}Co_{30}$) to 380 Oe for the sample with the highest cobalt content ($Fe_{20}Co_{80}$), and this may be ascribed to higher anisotropy energy of Co.

2.3. Oxide Coted Nickel-iron (NiFe) Nanoparticles

Herein, we report a facile polyol method for the preparation of high magnetization and nearly superparamagnetic NiFe nanoparticles without using any surfactant and deoxygenated conditions, while realizing the possibility for controlling the composition of the nanoparticles by varying some of ratio between iron and nickel precursors. The structure, composition, size, shape, surface coating, and magnetic properties of the synthesized nanoparticles were examined in detail and discussed the obtained results.

2.3.1. Experimental

2.3.1.1. Materials

High purity analytical reagent grade iron chloride tetrahydrate ($FeCl_2.4H_2O$), nickel chloride hexa hydrate ($NiCl_2.6H_2O$), polyethylene glycol (PEG), sodium hydroxide (NaOH) and ethyl alcohol were purchased from Sigma Aldrich and used in synthetic reaction without any further treatment.

2.3.1.2. Synthesis of NiFe Nanoparticles

NiFe nanoparticles were prepared with different chemical compositions such $Ni_{20}Fe_{80}$, $Ni_{50}Fe_{50}$, and $Ni_{80}Fe_{20}$. Firstly, in round bottom flasks, mixtures of ($FeCl_2.4H_2O$) and ($NiCl_2.6H_2O$) with a suitable amount of Polyethylene glycol (PEG) which was used as dispersing media was prepared in different compositions. For all the compositions, the pH of the solutions was adjusted in between 10-11 by adding NaOH prior to reduction process. Then the PEG-metal salts solution was heated up to 200°C for 30 minutes, then increase the temperature to 300°C, and was continuously stirred using a magnetic stirrer, refluxed at this temperature for 2 hours. Then the solutions cooled down to room temperature and washed several times using ethanol and water and then collected by using the magnet. The as-synthesized NiFe nanoparticles were annealed in presence of hydrogen for 2 hours.

2.3.2. Characterization

Structural characterizations were performed by X-ray diffraction (XRD) analysis using Rigaku RiNT 2200, which was carried out at the voltage of 40 kV and the current of 40 mA, by employing a scanning rate of 2°/min and a step size of 0.01 in the 2θ range from 20 to 80° with Cu K_α radiation. The size and morphology of the synthesized nanoparticles were characterized using Tecnai G2 F20 Scanning/Transmission electron microscope (S/TEM) with field emission system operated at 200 kV. Energy dispersive spectroscopy (EDS), which was mounted on S/TEM, was used for elemental analysis. The magnetic properties of the synthesized nanoparticles were measured by vibrating sample magnetometer (VSM) with an external magnetic field ranging from –13 kOe to +13 kOe.

2.3.3. Characterization

2.3.3.1. Crystal Structure

The X-ray diffraction patterns of the three compositions of NiFe nanoparticles which synthesized using different molar ratios of metal precursors are shown in Figure 24 (a, b, c). The peaks can be mainly indexed at the crystal planes of (111), (200), and (220) which match well with the face-centered cubic (fcc) structure of bulk NiFe alloy.

However, there are some peaks (marked by the asterisks in figure 24) for the composition with high iron content ($Ni_{20}Fe_{80}$) which cannot be indexed to fcc structure of NiFe. These diffraction peaks correspond well to those of spinel oxide ($(FeNi)_3O_4$) as reported by other group, where the high concentration of Fe precursor will result in the formation of oxide phases [80]. Furthermore, the oxide phase not only indicated from the XRD results but also clearly appear in the TEM images and also EDS analysis as we will be discuss later in the morphology part.

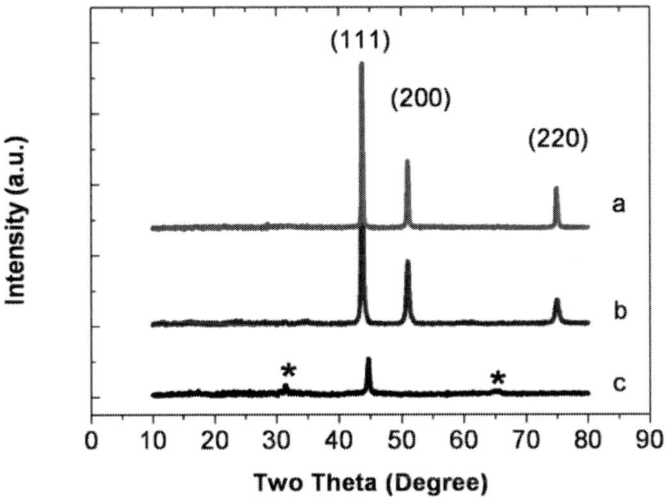

Figure 24. XRD patterns of different compositions of nickel ferrite (NiFe) NPs, (a) $Ni_{80}Fe_{20}$, (b) $Ni_{50}Fe_{50}$, and(c) $Ni_{20}Fe_{80}$ [91].

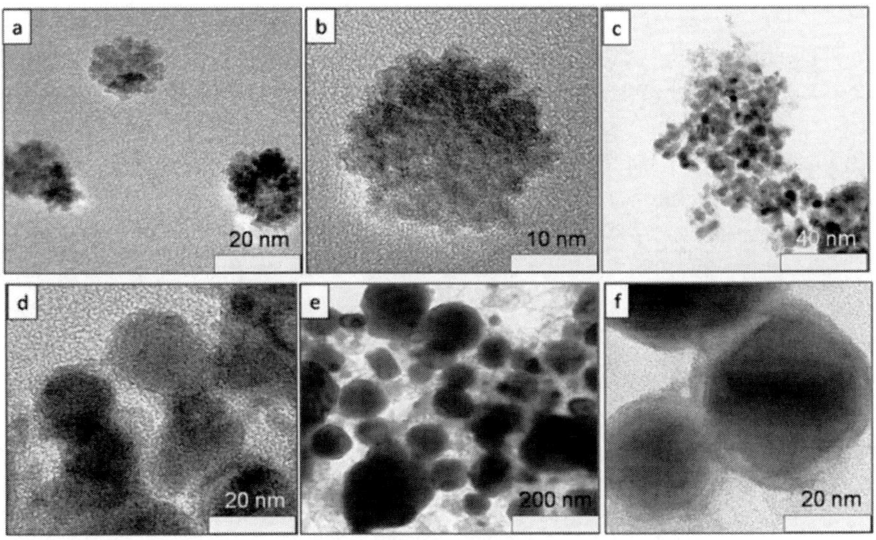

Figure 25. TEM images for NiFe nanoparticles (a, b and c) As-synthesized (d) Annealed NiFe nanoparticles, and (e and f) thin oxide layer [91].

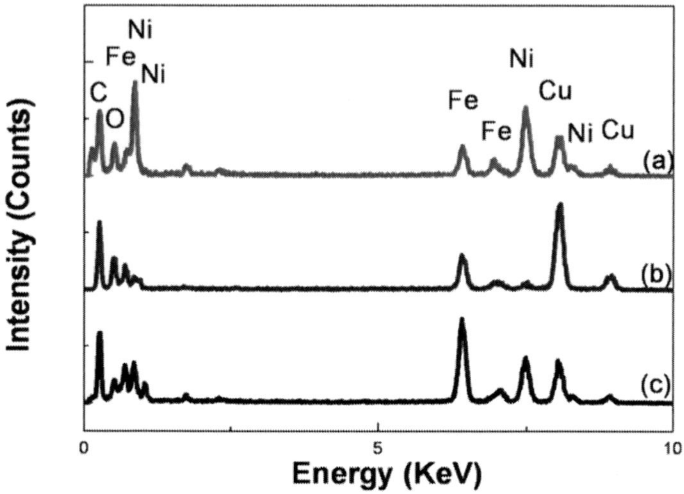

Figure 26. EDS patterns of (a) $Ni_{80}Fe_{20}$ (b) $Ni_{50}Fe_{50}$, and (c) $Ni_{20}Fe_{80}$ showing the presence of Fe and Ni and very small percent of oxygen elements in the NPs (Cu and C peaks are due to carbon copper grid) [91].

2.3.3.2. Morphology Characterization

The shape and size of the as-synthesized and annealed NiFe nanoparticles are observed in Figure 25 using TEM. Figure 25 (a, b, c) shows the typical TEM images of the as-prepared NiFe nanoparticles which reveal that the nanoparticles are monodisperse with 15 nm average size and have different shapes of flower-like and sphere shape depending on the composition of NiFe alloy (Figure 25b).

The such smaller size and excellent dispersion of the nanoparticles even we didn't used any surfactant or reducing agent in our reaction is mainly attributed to the using of polyethylene glycol (PEG). Hence, it acts not only as a solvent medium for the reaction but also as a reducing agent, where it is well known that the PEG is characterized by its strongest reducing properties, which lead to formation of a thin monolayer of the PEG (Figure 25d) on the surface of the nanoparticles not only to stabilize the particles in the solution but also to act as a protective agent to inhibit the particle growth [87, 104]. Furthermore, the condition used in our reaction of heating the PEG-metal salts solution firstly up to 200°C for 30 minutes instead of direct heating the solution to the refluxing temperature of 300ºC is important step, where it is helpful for permitting slowly growth of the nanoparticles. On the other hand, annealing the as-synthesized NiFe nanoparticles resulted in an increment in particle size to be in the range of 50-150 nm with still the monodispersity there

(Figure 25e). Furthermore, formation of oxide shell around the particle was observed (Figure 25f). This oxide shell is considered very important as a protective agent against oxidation. Same behaviors of oxide shell formation around the annealed nanoparticles within just exposure to the natural air have been reported with FeCo NPs and Fe NPs [87].

Further characterizations for the three different compositions of NiFe nanoparticles were obtained using EDS analysis which reveals that there are mainly two elements of Fe and Ni are detected, in addition to presence small percent of oxygen element which consider as an indicator about formation of the oxide shell around the nanoparticles. Meanwhile, we can see that an increase in the iron percent with increasing the iron on the composition and same with nickel metal. On the other hand, and especially in case of the bioapplications part we should mention that the NiFe nanoparticles without shells may, of course, display some toxic properties, but in our case the existence of a thin oxide layer around the NiFe nanoparticles, as evident from the TEM images in Figure 25f, and also the PEG layer with high bio-compatibility may significantly decrease the toxicity levels and thus can be termed as less harmful if used in drug delivery as one kind of its application.

2.3.3.3. Magnetic Properties

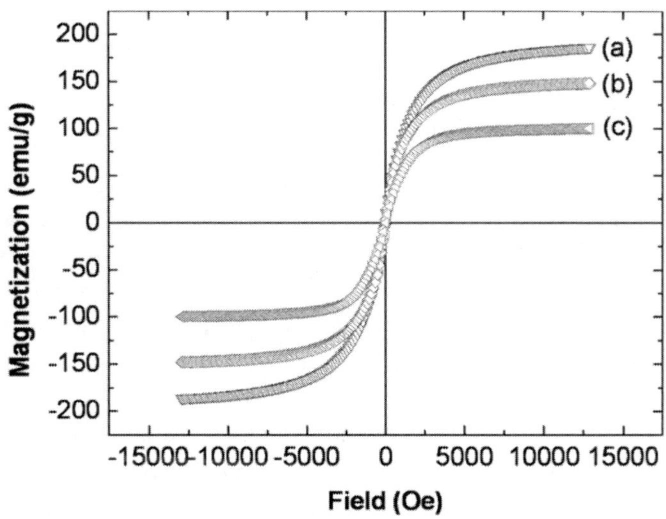

Figure 27. Magnetization curves of NiFe NPs for different compositions: (a) $Ni_{20}Fe_{80}$, (b) $Ni_{50}Fe_{50}$ and (c) $Ni_{80}Fe_{20}$ measured using VSM at room temperature (300 K) [91].

The magnetic properties of the NiFe nanoparticles were measured at room temperatures using VSM in an external magnetic field ranging from –13 kOe to +13 kOe. Figure 27 shows the magnetization curves for the three different compositions of $Ni_{80}Fe_{20}$, $Ni_{50}Fe_{50}$, and $Ni_{20}Fe_{80}$. Since, the saturation magnetization value of NiFe NPs depends on the ratio of Fe and Ni atoms, as expected, in both the cases, an increase in the saturation magnetization with iron content was observed due to the higher magnetic moments of Fe than Ni. The magnetization value increase from 99.6 emu/g for the composition of low iron content ($Ni_{80}Fe_{20}$) to 147.6 emu/g for the composition with equally amount of nickel and iron to reach maximum of 186.1 emu/g for $Ni_{20}Fe_{80}$. Up to our knowledge and based on the reported studied, the saturation magnetization value obtained here is consider the highest, and this higher value may be attributed to the larger particle size with negligible surface spin disorder. Moreover, the high crystallinity of the synthesized NiFe nanoparticles, as observed through the XRD patterns and TEM image (Figure 25e), also leads to a negligible surface spin canting and thus justifies the high value of magnetization. On the other hand, the coercivity (Hc) of the three compositions is decreased with nickel content from 57.66, 21, and 20 Oe for the $Ni_{20}Fe_{80}$, $Ni_{50}Fe_{50}$, and $Ni_{80}Fe_{20}$ respectively. The very small coercivity value here also is considered a very important parameter for the bioapplication as we mentioned above in the introduction part [108].

CONCLUSION

We have developed a simplistic sonochemical method to synthesize the Fe_3O_4 NPs which have been successfully synthesized using inexpensive and non-toxic metal salts as reactants. X-ray diffraction (XRD) and EDS patterns shows that there is an evident characteristic of magnetite phase (Fe_3O_4). It is also noteworthy that Transmission electron microscopy (TEM) measurements revealed that the as NPs prepared have size distribution in relatively narrow range and moderate monodispersity.

Uniform ferromagnetic magnetite nanocubes were synthesized using a cost effective, simple sonochemical method in the absence of any surfactant and by using nontoxic metal salt as reactant. Transmission electron microscopy (TEM) and scanning electron microscopy (SEM) show that the as-synthesized nanoparticles are cubic in shape with medium particle size of 80 nm. Annealing the sample at 400 °C and 600 °C resulted in an increment in particle size to be in the range of 80-85 and 85-90 nm, respectively with still the uniform cubic shape. On the other hand, further increase in annealing temperature to 700 °C leads to a distortion of the cube shape and broad size distribution to be in the range of 40-90 nm. The magnetic properties of the synthesized magnetite nanoparticles were measured by means of Physical Property Measurement System (PPMS) at different temperatures and the result shows a saturation magnetization of 91 emu/g at 5 K for the as-synthesized NPs. However, the magnetization was observed to be 94.8 emu/g at 5 K for 600 °C annealed sample. The magnetite phase (Fe_3O_4) was stable up to the annealing temperature of 600 °C, and small amount of maghemite phase of γ-Fe_2O_3 was formed at 700 °C annealing temperature.

Monodisperse $CoFe_2O_4$ NPs with uniform shape and size has been syntheiszed using rapid single-sonochemical reaction and without any kind of surfactant and deoxygenated protection or subsequent calcination process. Also, different morphological structures of spherical and cubic shape with different particle sizes were obtained by changing the synthesis solvent medium. Magnetization values of 71, 33.8, and 87 emu/g were obtained for the samples synthesized in aqueous medium, alcoholic, and mixed solution of water and ethanol in 1:1 volume ratio, respectively. Further, on adding PVP to the aqueous medium, the particles size has been observed to decrease with the corresponding decreases in the degree of crystallinity as well as in saturation magnetization to obtain the value of 70.5 emu/g.

We have developed a simplistic sono-chemical method to synthesize the $Fe_3O_4@SiO_2$ core-shell NPs which have been successfully synthesized using inexpensive and non-toxic metal salts as reactants. X-ray diffraction (XRD) and EDS analysis patterns shows that there is an evident characteristic of silica encapsulated magnetite phase ($Fe_3O_4@SiO_2$). It is also noteworthy that Transmission electron microscopy (TEM) measurements revealed that the as NPs prepared have size distribution in relatively narrow range and moderate monodispersity.

We also succeeded to synthesize high magnetization FeCo nanoparticles using facile modified polyol method without surfactant. The parameters of the reaction conditions, time, hydroxyl ion concentration and PEG were used while varying the quantities of Fe and Co precursors to effectively control the size and the magnetization value of FeCo nanoparticles. Transmission electron microscopy (TEM) shows that the synthesized nanoparticles are spherical in shape and monodisperse with average particle size of 10 nm before annealing. After the samples were annealed at 600 °C in presence of hydrogen, the particle sizes were increased to about 50 to 90 nm.

High magnetization, Monodisperse NiFe@oxide shell nanoparticles with controllable the compositions has also been synthesized using easy and cost effective polyol method. Transmission electron microscopy (TEM) shows that the synthesized nanoparticles are flower-like in shape with average particle size of 15 nm before annealing. After the samples were annealed in presence of hydrogen, the particle sizes were increased to be in-between 50 to 150 nm. Formation of oxide shell around the nanoparticles was confirmed by TEM images and EDS analysis.

These synthesized magnetite nanocubes and nanoparticles are expected to use in biofunctionalization due to their high surface to volume ratio and hydrophilic properties in addition to their high magnetic moment. In view of the obtained results of high crystalline structure, high magnetization value and good dispersion properties, we consider our $CoFe_2O_4$ NPs carry enough significance for use in various bioapplications. We believe that this Fe_3O_4@SiO_2 core-shell NPs can be used for bio and as well as many further applications. The synthesized nanoparticles (FeCo and NiFe) are expected to be suitable as labels for different biomedical application due to their high saturation magnetization and very small coercivity. Moreover, these synthesis method are considered a promising for uncomplicated synthesis of different kinds of nanoparticles.

REFERENCES

[1] Deliyanni, E. A., Bakoyannakis, D. N., Zouboulis, A. I. & Matis, K. A. *Chemosphere*. 2003, 50, 155-163.

[2] Yamada, M., Waki, I., Sakairi, M., Sakamoto, M. & Imai, T. *Chemosphere*. 2004, 54, 1475-1480.

[3] Trivedi, P., & Axe, L. *Environ. Sci. Technol*. 2000, 34, 2215-2223.

[4] Benjamin, M. M. & Leckie, J. O. *J. Colloid Interface Sci*. 1981, 79, 209-221.

[5] Yean, S., Cong, L., Yavuz, C. T., Mayo, J. T., Yu, W. W., Kan, A. T., Colvin, V. L. & Tomson, M. B. *J. Mater. Res*. 2005, 20, 3255-3264.

[6] Jendelova, P., Herynek, V., Urdzikova, L., Glogarova, K., Kroupova, J., Andersson, B., Bryja, V., Burian, M., Hajek, M. & Sykova, E. J. *Neurosci. Res*. 2004, 76, 232-243.

[7] Wiltshire, M. C. K., Pendry, J. B., Young, I. R., Larkman, D. J., Gilderdale, D. J. & Hajnal, J. V. *Science*. 2001, 291, 849-851.

[8] Loubeyre, P., Jaegere, T. D., Bosmans, H., Miao, Y., Ni, Y., Landuyt, W. & Marchal, G. J. *Mag. Reson. Imaging*. 1999, 9, 447-453.

[9] Cheng, F. Y., Su, C. H., Yang, Y. S., Yeh, C. S., Tsai, C. Y., Wu, C. L., Wu, M. T. & Shieh, D. B. *Biomaterials* 2005, 26, 729-738.

[10] Song, H. T., Choi, J. S., Huh, Y. M., Kim, S., Jun, Y. W., Suh, J. S. & Cheon, J. *J. Am. Chem. Soc*. 2005, 127, 9992-9993.

[11] Bucak, S., Jones, D. A., Laibinis, P. E. & Hatton, T. A. *Biotechnol. Prog.* 2003, 19, 477-484.

[12] Hatch, G. P. & Steler, R. E. *J. Magn. Magn. Mater*. 2001, 225, 262-276.

[13] Xie, X., Zhang, X., Zhang, H., Chend, D. & Fei, W. *J. Magn. Magn. Mater.* 2004, 277, 16-23.

[14] Nam, J.-M., Thaxton, C. S. & Mirkin, C. A. *Science* 2003, 301, 1884-1886.

[15] Shinkai, M. & Ito, A. *Adv. Biochem. Eng. Biotechnol*. 2004, 91, 191-220.

[16] Cornell, R. M. & Schwertmann, U. The Iron Oxides, VCH, New York, 1996.

[17] Lalatonne, Y., Richardi, J. & Pileni, M. P. *Nat. Mater*. 2004, 3, 121-125.

[18] Redl, F. X., Cho, K. S., Murray, C. B. & O'Brien, S. *Nature*. 2003, 423, 968-971.

[19] Bao, S. J. Zhang, X. G. & Liu, X. M. *Fenzi Cuihua*. 2003, 17, 96-100.

[20] Moser, A., Takano, K., Margulies, D. T., Albrecht, M. Sonobe, Y., Ikeda, Y., Sun S. & Fullerton, E. E. *J. Phys. D: Appl. Phys*. 2002, 35, R157- R167.

[21] Matsushita, N., Chong, C. P., Mizutani, T. & Abe, M. *J. Appl. Phys.* 2002, 91, 7376-7378.

[22] Sorenson, T. A., Morton, S. A., Waddill, G. D. & Switzer, J. A. *J. Am. Chem. Soc.* 2002, 124, 7604-7609.

[23] Black, C. T, Murray, C. B., Sandstrom, R. L. & Sun, S. *Science*. 2000, 290, 1131-1134.

[24] Zeng, H., Li, J., Liu, J. P., Wang, Z. L. & Sun, S. *Nature*. 2002, 420, 395-398.

[25] Lu, Z. L., Zou, W. Q., Lv, L. Y., Liu, X. C., Li, S. D., Zhu, J. M., Zhang, F. M. & Du, Y. W. *J. Phys. Chem*. 2006, B 110, 23817-23820.

[26] Liao, Z. M., Li, Y. D., Xu, J., Zhang, J. M., Xia, K. & Yu D-P. *Nano Lett.* 2006, 6, 1087-1091.

[27] Zhang, D., Liu, Z., Han, S., Li, C., Lei, B., Stewart, M. P., Tour, J. M., & Zhou, C. *Nano Lett*. 2004, 4, 2151 – 2155.

[28] Shinkai, M. & Ito, A. *Adv. Biochem. Eng. Biotechnol.* 2004, 91, 191-220.

[29] Oliveira, L. C. A., Rios, R. V. R. A., Fabris, J. D., Garg, V., Sapag, K. & Lago, R. M. *Carbon* 2002, 40, 2177–2183.

[30] Wu, R., Qu, J. & Chen, Y. *Wat. Res.* 2005, 39, 630-638.

[31] Lee, Y., Lee, J., Bae, C., Park, J., Noh, H., Park, J. & Hyeon, T. *Adv. Funct. Mater.* 2005, 15, 503-509.

[32] Yang, H., Ogawa, T., Hasegawa, D. & Takahashi, M. *J. Appl. Phys.* 2008, 103, 526-528.

[33] Schmidt-Nielsen, K. Scaling: Why is Animal Size so Important? New York, NY: Cambridge University Press. 1984.

[34] Xiong, Y., Ye, J., Gu, X. & Chen, Q. W. *J. Phys. Chem.* C. 2007, 111, 6998-7003.

[35] Gao, G., Liu, X., Shi, R., Zhou, K., Shi, Y., Ma, R., Eiji, T. M. & Qiu, G. *Cryst. Growth. Des.* 2010, 10, 2888-2894.

[36] Pauly, M., Pichon, B., Albouy, P., Fleutot, S., Leuvery, C., Trassin, M., Gallani, J. & Colin, S. *J. Mater. Chem.* 2011, 21, 16018-16027.

[37] Kim, D., Lee, N., Park, M., Kim, B. H., An, K. & Hyeon, T. *J. Am. Chem. Soc.* 2009, 131, 454-455.

[38] Suslick, K. S. *Science.* 1990, 247, 1439-1445.

[39] Bang, J. H. & Suslick, K. S. *Adv. Mater.* 2010, 22, 1039-1059.

[40] Kumar, R. V., Koltypin, Y., Cohen, Y. S., Cohen, Y., Aurbach, D., Palchik, O., Felner, I. & Gedanken, A. *J. Mater. Chem.* 2000, 10, 1125-1129.

[41] Marchegiani, G., Imperatori, P., Mari, A., Pilloni, L., Chiolerio, A., Allia, P., Tiberto, P., & Suber, L. *Ultrasonics Sonochemistry.* 2012, 19, 877–882.

[42] Cheng, J. P., Ma, R., Shi, D., Liu, F. & Zhang, X. B. *Ultrasonics Sonochemistry.* 2011, 18, 1038-1042.

[43] Goh, S. C., Chia, C. H., Zakaria, S., Yusoff, M., Haw, C. Y., Ahmadi, S. H., Huang, N. M. & Lim, H. N. *Mater. Chem. Phys.* 2010, 120, 31-35.

[44] Lee, J. G., Park, J. Y. & Kim, C. S. *J. Mater. Sci.* 1998, 33, 513-518.

[45] Rao, G. S. N., Caltun, O. F., Rao, K. H., Rao, P. S. V. S. & Rao, B. P. *J. Magn. Magn. Mater.* 2013, 341, 60-64.

[46] Baldi, G., Bonacchi, D. & Innocenti, C. *J. Magn. Magn. Mater.* 2007, 311, 10-16.

[47] Shultz, M. D., Calvin, S., Fatouros, P. P., Morrison, S. A. & Carpenter, E. E. *J. Magn. Magn. Mater.* 2007, 311, 464-468.

[48] Schmid, G. Nanoparticles: from theory to application, Weinheim, Wiley-VCH, 2004.

[49] Mazario, E., Herrasti, P. Morales, M. P. & Menendez, N. *Nanotechnology*. 2012, 23, 355708.

[50] Zi, Z., Sun, Y., Zhu, X., Yang, Z., Dai, J. & Song, W. *J. Magn. Magn. Mater.* 2009, 321, 1251-1255.

[51] Gul, I.H. & Maqsood, A. *J. Alloy Compound*. 2008, 465, 227-231.

[52] Suslick, K.S. *Science*. 1990, 247, 1439-1445.

[53] Dang, F., Enomoto, N., Hojo, J. & Enpuku, K. *Ultrason. Sonochem.* 2009, 16, 649-654.

[54] Bang, J. H. & Suslick, K. S. *Adv. Mater*. 2010, 22, 1039-1059.

[55] Chen, Q., Boothroyd, C., Tan, G. H., Sutanto, N., Soutar, A. M. & Zeng, X. T. *Langmuir*. 2008, 24, 650-653.

[56] Morel, A. L., Nikitenko, S. I., Gionnet, K., Wattiaux, A., Him, J. L. K., Labrugere, C., Chevalier, B., Deleris, G., Petibois, C., Brisson, A. & Simonoff, M. *acsnano*. 2008, 2(5), 847-856.

[57] Li, L., Choo, E. S. G., Yi, J., Ding, J. Tang, X. & Xue, *J. Chem. Mater.* 2008, 20, 6292-6294.

[58] Dong, K. Y., Selvan, S. T. Lee, S. S. Georgia. C., Papaefthymiou, G. C., Kundaliya, D. & Ying, J. Y. *J. Am. Chem. Soc.* 2005, 127, 4990-4991.

[59] Sun, S., Murray, C.B., Weller, D., Folks, L. & Moser, A. *Science*. 2000, 287, 1989-1992.

[60] Lacheisserie, E. T. & Gignoux, D, Schlenker M. Magnetism: Materials and Spplications; Springer, 2004.

[61] Bader, S. D. Colloquium: Opportunities in nanomagnetism; *Rev. Mod. Phys.* 2006, 78, 1-15.

[62] Lu, A. H., Salabas, E. L. & Schüth, F. Functionalization, and application; *Angew. Chem. Int.* Ed. 2007, 46, 1222-1244.

[63] Pankhurst, Q.A., Connolly, J. Jones, S.K. & Dobson, J. *J. Phys. D: Appl. Phys.* 2003, 36, R167-R181.

[64] Furlani, E. P. & Ng, K. C. *Phy. Rev.* E. 2006, 73, 061919-061919-3.

[65] Joseyphus, R. J., Shinoda, K., Kodama, D. & Jeyadevan, B. *Mater. Chem. and phys.* 2010, 123, 487-493.

[66] Wang, C., Peng, S., Lacroix, L. M. & Sun, S. *Nano Res.* 2009, 2, 380-385..

[67] Nguyen, Q., Chinnasamy, C. N., Yoon, S. D., Sivasubramanian, S., Sakai, T., Baraskar, A., Mukerjee, S., Vittoria, C. & Harris, V. G. *J. Appl. Phys.* 2008, 103, 07D532.

[68] Tzitzios, V., Basina, G., Niarchos, D., Wanfeng, Li. & Hadjipanayis, G. *J. Appl. Phys.* 2011, 109, 07A313-07A313-3.

[69] Zábranskýa, K., Davida, B., Pizúrováa, N., Schneeweissa, O., Zborilb, R. & Mašlán, M. *AIP Conf. Proc.* 2010, 102, 1258.

[70] Z′abransk′y, K. & Schneeweiss, O. *Journal of Electrical Engineering.* 2010, 61(5), 299–301.

[71] Chaubey, G. S., Barcena, C., Poudyal, N., Rong, C., Gao, J., Sun, S. & Liu, J. P. *J. Am. Chem. Soc.* 2007, 129, 7214-7215.

[72] Li, H. F. & Ramanujan, R. V. *Trans. Indian Inst. Met.* 2005, 58(6), 965-970.

[73] Kim, C. W., Kim, Y. H., Cha, H. G., Lee, D. K. & Kang, Y. S. J Nanosci Nanotechnol. 2006, 6(11), 3417-3421.

[74] Sun, C., Lee, J. S. H. & Zhang, M. *Adv. Dru. Del. Rev.* 2008, 60, 1252-1265.

[75] Fievet, F., Lagier, J. P. & Figlarz, M. *MRS Bull.* 1989, 14, 29-34.

[76] Suresh, G., Saravanan, P., & Babu D. R. *J. of. Nan. Research.* 2011, 15, 21-28.

[77] Patnaik, P. Wiley, New Jersey; Ch. 34, 2007.

[78] Glavee, G. N., Klabunde, K. J., Sorensen, C. M. & Hadjipanayis, G. C. *Inorg. Chem.* 1995, 34(1), 28-35.

[79] Chen, Y., Luo, X. Yue, G. H. Luo, X. & Peng, D. L. *Mater. Chem. Phys.* 2009, 113, 412-416.

[80] Chen, Y., Luo, X., Yue, G. H., Luo, X. & Peng, D. L. *Mater. Chem. Phys.* 2009, 113, 412-416.

[81] Ban, I., Drofenik, M. & Makovec, D. J. Mag. Mag. Mater. 2006, 307, 250-256.

[82] Suh, Y. J., Jang, H. D., Chang, H., Kim, W. B. & Kim, H. C. *Powder Technol.* 2006, 161, 196-201.

[83] Wu, H.Q., Cao, Y. J., Yuan, P. S., Xu, H. Y. & Wei, X. W. *Chem. Phys. Lett.* 2005, 406, 148-153.

[84] Cushing, B. L., Golub, V. & O'Connor, C. J. *J. Phys. Chem. Solids.* 2004, 65, 825-829.

[85] Liao, Q. Tannenbaum, R. & Wang, Z. L. *J. Phys. Chem. B.* 2006, 110(29), 14262-14265.

[86] Shafi, K. V. P. M., Gedanken, A., Goldfarb, R. B. & Felner, I. *J. Appl. Phys.* 1997, 81, 6901-6905.

[87] Abbas, M., Islam, M. N., Rao, B. P., Ogawa, T. Takahashi, M. & Kim, C. G. *Materials Letts.* 2013, 91, 326-329.

[88] Islam, M. N., Phong, L. V., Jeong, J.-R. & Kim, C. G. *Thin Solid Film.* 2011, 519, 8277-8279.

[89] Abbas, M., Takahashi, M. & Kim, C. G. *J. Nanopart Res*. 2013, 15, 1354-1366.

[90] Abbas, M., Rao, B. P., Islam, M. N., Kim, K. W. Naga, S.M., Takahashi, M. & Kim, C. G. *Ceramics International*. 2014, 40, 3269–3276.

[91] Islam, M. N., Abbas, M. & Kim, C. G. *Current Applied Physics.* 2013, 13, 2010-2013.

[92] Makino, K., Mossoba, M. M., Riesz, P. *J Am Chem Soc.* 1982, 104, 3537-3539.

[93] Sidhu, P. S., Gilkes, R. J., Posner, A. M. *Journal of Inorganic & Nuclear Chemistry.* 1977, 39 (11), 1953-1958.

[94] Kim, K. C., Kim, Y. S. *J Magn.* 2010, 15(4), 179-184.

[95] Dunlop D, J. *J. Geophys. Res*. 1973, 78, 1780-1793.

[96] Enkin, R. J. and Dunlop, D. J. *J Geophys. Res.* 1987, 92, 12726-12740.

[97] Kodama, R. H., Berkowitz, A. E., McNiff, E. J. & Foner, S. *Phys. Rev. Lett.* 1996, 77, 394-397.

[98] Roy, S., Dubenko, I., Edorth, D. D. & Ali, N. *J. Appl. Phys.* 2004, 96, 1202- 1208.

[99] Rajendran, M., Pullar, R. C., Bhattacharya, A. K., Das, D., Chintalapudi, S, N. & Majumdar, C. K. *J. Magn. Magn. Mater.* 2001, 232, 71-83.

[100] Patil, R. P., Hankare, P. P., Garadkar, K. M. & Sasikala, R. *J. of Alloy and Compounds.* 2012, 523, 66-71.

[101] Goss, C. *J. Phys. Chem. Miner.* 1988, 16,164-171.

[102] Kneller, E. F. & Luborsky, F. E. *J. Appl. Phys.* 1963, 34, 656-659.

[103] Narayanan, R. & El-Sayed, M. A. *J. Am. Chem. Soc.* 2003, 125, 8340-8347.

[104] Abbas, M., Rao, B. P., Naga, S. M., Takahashi, M. & Kim, C. G. *Ceramics International*. 2013, 39, 7605–7611.

[105] Kolhatkar, A. G., Jamison, A. C., Litvinov, D., Willson, R. C. & Lee, T. R. *Int. J. Mol. Sci*. 2013, 14, 15977-16009.

[106] Song, Q. & Zhang, Z. J. J. Am. Chem. Soc. 2004, 126, 6164–6168.

[107] Islam, M. N., Abbas, M., Sinha, B., Joeng, J. -R. and Kim, C. G. *Electron. Mater. Lett.* 2013, 9(6), 817-820

[108] Roca, A. G., Niznansky, D., Poltierova, V. J., Bittova, B., Fernández, G. M. A., Serna, C. J. & Morale, M. P. *J. Appl. Phys.* 2009, 105, 114309(1-7).

In: Magnetic Nanoparticles
Editor: Nora P. Sabbas
ISBN: 978-1-63117-434-6

Chapter 6

ELECTROSPUN MAGNETOACTIVE FIBROUS NANOCOMPOSITES: FABRICATION AND APPLICATIONS IN BIOMEDICINE

Ioanna Savva and Theodora Krasia-Christoforou
University of Cyprus, Department of Mechanical
and Manufacturing Engineering, Nicosia, Cyprus

ABSTRACT

Materials capable of responding to external stimuli including pH, temperature, magnetic and electric field, undergoing conformational changes are considered to be one of the most exciting and emerging classes of advanced materials receiving considerable scientific interest especially in the biomedical field. Stimuli-responsive polymers in the form of micro- or nanofibers have received great attention during the last 10 years and have been exploited in a diverse range of biomedical applications such as drug delivery, tissue engineering, bioseparation and biosensing. One of the most popular and versatile fiber fabrication methods used for the production of fibers is electrospinning. Its simplicity, cost-effectiveness and applicability not only to pristine synthetic and natural polymers but also to composites, enables the development of polymer-based fibrous nanocomposites *via* the combination of polymers with inorganic nanofillers. Among such nanoadditives, magnetic nanoparticles capable of interacting with an externally applied magnetic field, are particularly attractive owing to their potential biomedical applications including magnetically-triggered drug delivery, magnetic cell seeding, magnetic bioseparation, hyperthermia cancer treatment and contrast enhancement in magnetic resonance imaging. In this chapter, an introductory section on electrospinning and on different parameters influencing this process is initially provided. The different fabrication routes for generating electrospun magnetoactive polymer-based (nano)fibrous materials are briefly discussed and finally the applicability of these materials in the biomedical field including tissue engineering, drug delivery, hypethermia treatment and biosensing is reviewed.

OVERVIEW OF THE ELECTROSPINNING PROCESS

The history of electrospinning began back in 1900 when John Francus Cooley filed the 1st electrospinning patent [1]. However, only in early 2000s scientific publications on electrospinning started appearing, whereas from that time onwards an exponential increase in the number of publications on electrospun nanofibers has been observed, owned to the versatility of this field in the development of polymer-based advanced materials with unique properties and consequently potential technological applications. Figure 1 shows the number of publications per year on "*Nanofibers and Electrospinning*" (Source: Web of Science, December 2013) demonstrating the enormous interest by the scientific community for this highly promising nanofiber fabrication technique.

Although currently there are several methods that can be used to prepare one dimensional (1D) nanomaterials namely drawing [2], self-assembly [3, 4], melt-blow [5], phase-separation [4] and template synthesis [6], the most popular and versatile fiber fabrication method used for the production of synthetically or naturally-derived fibrous materials is electrospinning [7]. Electrospinning can be used to produce continuous polymer (nano)fibers [8, 9, 10] with diameters ranging from micrometers (10-100 μm) down to a few nanometers (10-100 nm) [8] and lengths up to meters [11]. Its simplicity, cost-effectiveness and applicability not only to polymers [12, 8] but also to composites, enables the development of polymer-based fibrous nanocomposites *via* the combination of polymers with inorganic nanofillers [13].

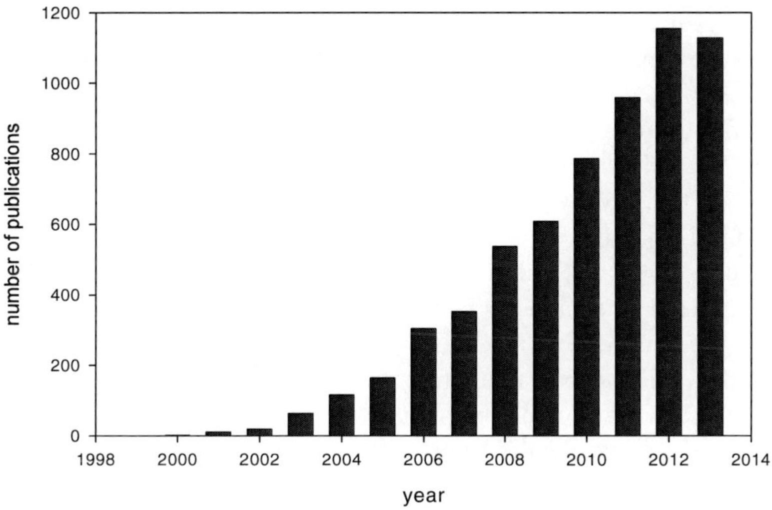

Source: Web of Science, December 2013.

Figure 1. Number of publications on nanofibers and electrospinning per year.

Basic Principles

A simple electrospinning set-up consists of four major components: A high-voltage power supply, a syringe with a metallic needle (containing the polymer solution or melt), a syringe pump used for delivering the solution through the spinneret at a constant and controllable rate and a grounded conductive collector on which the produced fibers are collected (Figure 2).

Direct current (DC) power supplies are usually used for electrospinning although the use of alternating current (AC) potential is also feasible [14]. A high-voltage power supply is connected with both the needle and the collector, whereas the positive electrode (anode) of the power supply is connected to the needle and the negative (cathode) is grounded to the collector [15]. In a typical fiber-generating process, a syringe is filled with a polymer solution and a high voltage (up to 30 kV) [16, 14] is applied between the syringe nozzle and a collector. The syringe may be placed perpendicularly, letting the polymer fluid drop with the help of gravity, or horizontally in respect to the grounded collector [17].

When a high voltage is applied, the pendent drop of the polymer solution at the nozzle of the spinneret is electrically charged and the induced charges are evenly distributed over the droplet surface. The electrostatic repulsion forces developed between the surface charges and the columbic force (electrostatic force) exerted by the external electric field are the major electrostatic forces on the droplet. Under these electrostatic interactions a conical fluid structure, called the Taylor cone, is formed at the tip of the syringe. At a critical voltage (i.e., once the strength of the electric field has surpassed a threshold value) the electrostatic repulsive forces overcome the surface tension of the solution and force a jet to erupt from the tip of the Taylor cone. If the applied voltage is not high enough, the jet will break up into droplets, a phenomenon called Rayleigh instability. If the voltage is sufficiently high, a stable jet will form near the tip of the Taylor cone [18]. The electrified jet is only stable near the tip of spinneret, after which the jet undergoes a stretching and whipping process. In solution electrospinning, as the charged jet accelerates toward lower-potential regions, the solvent evaporates whereas the resulting increase in the electrostatic repulsion of the charged polymer causes the fibers to elongate and reduce their diameters [19, 18]. With the solvent evaporation, the jet eventually becomes solidified on the collector in the form of randomly oriented (nano) fibers [20, 21, 15, 18, 19].

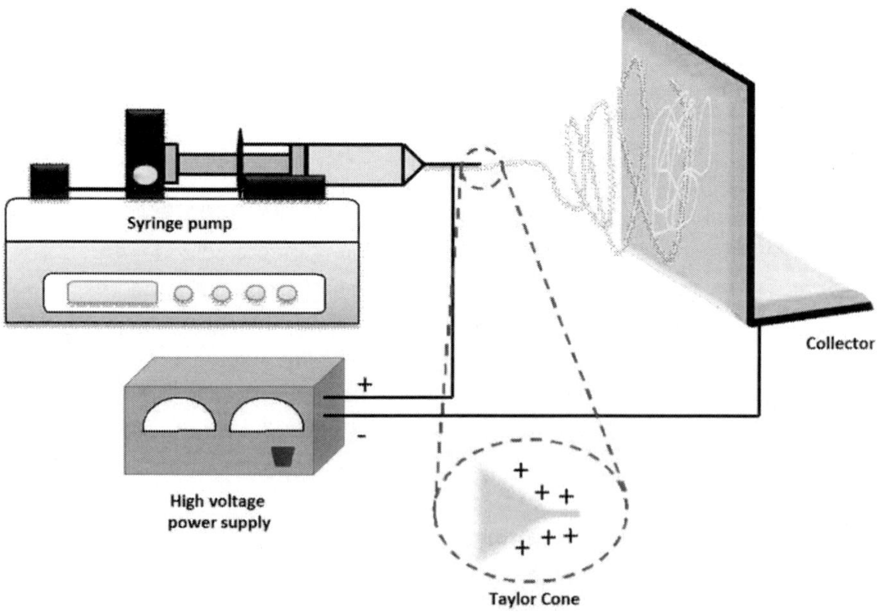

Figure 2. Schematic presentation of a basic electrospinning set-up.

Influencing parameters in the electrospinning process		
Solution properties	**Process parameters**	**Ambient parameters**
Viscosity/Concentration Conductivity/Solution charge density Surface tension Polymer molecular weight Dipole moment and dielectric constant	Applied voltage Polymer flow rate Needle-to-collector distance Collector type	Temperature Humidity

Figure 3. Parameters influencing the electrospinning process: (i) Solution properties, (ii) process and (iii) ambient parameters.

Influencing Parameters

In polymer nanofibers deriving from electrospinning, it is desirable that the fiber diameters are consistent and controllable, the fiber surface is defect-free, and continuous single nanofibers are collectable. However, research carried out throughout the years has shown that the above-mentioned targets are by no means easily achievable [19].

Many parameters may influence the transformation of a polymer solution into nanofibers through electrospinning, which are classified in terms of the solution properties [22, 23, 24, 19], process parameters [9, 19] and ambient parameters [25, 9, 22, 24]. Figure 3 summarizes the three influencing categories on the electrospinning process.

Based on those parameters, dissimilar results may be obtained using the same polymer and electrospinning set-up. Thus, it is difficult to give quantitative relationships that can be applied across a broad range of polymer/solvent systems. However, there are general trends which are useful when determining the optimum conditions for a certain system.

Solution Properties

Viscosity/**Concentration**

Solution viscosity which is governed by the polymer solution concentration strongly influences the dimensions and morphologies of electrospun polymeric fibers [26]. The solution concentration must be high enough for allowing for chain entanglement effects to take place. If the solution is too diluted then the polymer fiber will break up into droplets before reaching the collector due to surface tension effects. Consequently, at low polymer concentrations, the generation of beads and droplets has been observed. On the other hand, in too concentrated solutions the solution viscosity is very high thus hindering the solution flow rate through the capillary [27, 28, 22, 25].

Conductivity/Solution Charge Density

Solution conductivity also influences the morphology and diameters of the resulting fibers. An increased conductivity enhances the charge density at the surface of the jet and the charge carrying capacity; as a consequence the jet is subjected to a greater tensile force in the presence of an electric field [29, 30, 31], thus reducing the tendency of droplet formation during the process. Experimental investigations have demonstrated that an increase in solution conductivity results to a substantial decrease in nanofiber diameter. Furthermore it has been shown that the radius of the electrospun nanofiber jet is inversely related to the cube root of the electrical conductivity of the solution [32, 29].

Surface Tension

The impact of surface tension on the morphology and dimensions of electrospun fibers has also been investigated [22]. Surface tension plays a critical role in the electrospinning process since – based on different studies – it has been demonstrated that by reducing the surface tension of a nanofiber solution, bead-free fibers can be obtained. The use of different solvents influences the solution surface tension. Generally, solutions characterized by a high surface tension lead to the generation of sprayed droplets owing to jet instability [33]. Basically, surface tension determines the upper and lower boundaries of the electrospinning window if all other variables are held constant [34]. Moreover, the addition of salts to increase the net charge density or use of solvents with lower surface tension could also lead to the effective elimination of beads and to the generation of fibers with larger diameters [25].

Polymer Molecular Weight

The polymer molecular weight (MW) influences considerably the rheological and electrical properties of a solution such as viscosity, surface tension, conductivity and dielectric strength [35]. Researchers have been examining the relationship existing between MW and the morphological and dimensional characteristics of electrospun fibers [36]. From such studies it has been concluded that an increase in MW leads to the reduction in the number of beads and droplets generated during electrospinning. Additionally, polymers characterized by a narrow molecular weight distribution (MWD) led to the formation of uniform fibers at a lower concentration compared to solutions prepared using a broad MWD polymer of the same chemical composition. Moreover, it was found that the diameters of the electrospun fibers depended strongly on the MW of the polymer [22, 35], i.e., when using a low MW polymer, there is a tendency toward the formation of beaded structures, whereas upon increasing the MW continuous fibers are generated. In the latter case, the fiber diameter increases upon increasing the polymer MW [34, 35].

Dipole Moment and Dielectric Constant

Very few studies so far have carried out an in-depth investigation of the effect of the dipole moment and the dielectric constant on the morphology of the fibers deriving from electrospinning. In one example, polystyrene (PS) was electrospun into fibers starting from polymer solutions in eighteen different solvents [37]. The solvents with high dipole moment values were those resulting to the successful fabrication of PS electrospun fibers. Moreover, the number of PS fibers produced per unit time was found to correlate to the dipole moment and the dielectric constant [38, 22, 37].

Process Parameters

Applied Voltage

The strength of the applied electric field considerably influences the electrospinning process. Many studies show that by increasing the electric field strength the bead density is decreased, while a further increase leads to an increase in the average fiber diameter [39, 24]. In contrast, the formation of beads may occur upon increasing the applied voltage, while the jet viscosity increases and the solution is removed from the tip more quickly. The Taylor cone shape oscillates and becomes asymmetrical when the volume of the droplet on the tip becomes smaller [16]. Many authors evident that there is an optimal electric field strength range for a certain polymer/solvent system, as either too weak or too strong field will lead to the formation of beaded fibers [24, 39].

Polymer Flow Rate

The polymer flow rate has also an impact on the dimensions and geometry of the produced electrospun fibers as well as on the porosity of the fibrous material. Both, the fiber diameter and the pore size increase with increasing flow rate. Additionally, at high flow rates a significant amount of beaded defects can be observed, due to the incapability of fibers to dry completely before reaching the collector. Incomplete fiber drying also leads to the formation of ribbon-like fibers as compared to fibers with a circular cross section [28]. In general, it was found that lower flow rates yielded continuous fibers with smaller diameters [40, 22].

Needle-to-Collector Distance

Although the distance between the needle-tip and the collector is not a parameter that is influencing primarily the electrospinning process, it can still influence the fiber morphology and size. By increasing the distance from the Taylor cone, the fiber diameter decreases, whereas the formation of a beaded morphology is usually obtained by shortening the distance between the needle-tip and the collector. The latter may be attributed to inadequate drying of the polymer fiber prior to reaching the collector; consequently, a minimum distance is required to allow the fibers sufficient time to dry before reaching the collector [41, 42, 9, 28]. Finally, a decrease in the needle-to-collector distance results to the formation of shorter fibers [43].

Collector Type

The modification of collectors for fiber direction control has been widely investigated. Different geometries of the conductive collectors have been successfully developed including a cylindrical collector with high rotating speed, a cylindrical collector based on a pair of split electrodes, a thin wheel collector with sharp edge, and a frame collector [1, 19, 25]. It is noteworthy to mention that the more conductive collectors dissipated the charge of the fibers, whereas when this charge was not dissipated (i.e., in the cases where non-conductive collectors are used), the fibers repelled one another, resulting to the generation of a more porous structure [44]. Furthermore, porous collectors, such as paper and copper mesh, led to the production of less-packed structures compared to fibers collected on aluminum foil or water [22].

Ambient Parameters

To date, a limited number of studies examining the effects of ambient parameters such as temperature and humidity on the electrospinning process exist. As far as the temperature effect is concerned, it has been demonstrated that by increasing the temperature, fibers with a smaller diameter are obtained due to the decrease in the polymer solution viscosity at elevated temperatures. Moreover, an increase in the humidity level, results to the appearance of small circular pores on the surface of the fibers while a further increase leads to pore coalescing [22].

FABRICATION OF ELECTROSPUN MAGNETOACTIVE FIBROUS NANOCOMPOSITES

Synthetic materials capable of responding to external stimuli are one of the most exciting and emerging class of advanced materials receiving considerable scientific interest due to their potential applications in many scientific and technological fields [45]. In general, stimuli-responsive materials, the so called "smart" materials can be externally triggered upon variation of the solution pH [46] or the temperature [47, 48], *via* irradiation [49] and by applying a magnetic [50] or an electric [51] field, undergoing conformational changes.

Stimuli-responsive polymers in the form of micro- or nanofibers have received great attention during the last 10 years and have been exploited in a diverse range of biomedical applications such as drug delivery, tissue engineering, bioseparation and biosensing [52, 15].

Besides being a simple processing technique for fabricating (nano)fibers, electrospinning allows the co-processing of polymer mixtures as well as chemical cross-linking to take place, providing a variety of pathways for manipulating the chemical composition of the fibers. The above-mentioned, offer useful properties to the resulting fibrous materials including mechanical strength, elasticity, porosity and charged surface areas [20].

Moreover electrospinning enables the incorporation of a variety of functional micro/nano particles within the fibers [53]. Different nanoparticles or nanofiller types may be dispersed in polymer solutions, which are then electrospun to generate composites in the form of continuous (nano)fibers and (nano)fibrous assemblies. All these endow electrospinning with outstanding manufacturing capabilities whilst utilizing an easy and highly flexible process [17, 32].

Among such nanoadditives, magnetic nanoparticles capable of interacting with an externally applied magnetic field are particularly attractive. Magnetoresponsive polymer-based fibrous nanocomposites belonging to the broad category of stimuli-responsive materials, is a relatively new class of "soft" composite materials, consisting of magnetic nanoparticles embedded within a polymeric fibrous matrix. The presence of an externally applied magnetic field influences the properties of these materials rendering them useful in numerous technological and biomedical applications including magnetically-triggered drug delivery, magnetic cell seeding, magnetic bioseparation, hyperthermia cancer treatment, contrast enhancement agents in magnetic resonance imaging and biosensing.

To date numerous examples appear in the literature dealing with the fabrication of electrospun magnetoactive polymer-based fibrous (nano)composites that are mainly focused on the investigation of their morphological characteristics and magnetic properties. In such materials different types of polymers including polyurethanes, polymethacrylates, polyvinylpyrrolidone (PVP), polyacrylonitrile (PAN), polyvinyl alcohol (PVA), degradable polyesters such as poly(L-lactide) (PLLA), poly(L-lactide)-*co*-poly(glycolide) (PLGA) and polycaprolactone (PCL), as well as natural polymers and biopolymers have been combined with magnetic nanoparticles including Fe_3O_4, γ-Fe_2O_3, Co, Ni, FePt etc. In Table 1 the literature references on electrospun magnetoactive polymer-based materials are summarized and are categorized based on the polymer type and the kind of the magnetic nanoparticles embedded within the electrospun polymer (nano)fibers.

Different strategies have been employed to generate electrospun magnetoactive polymer-based (nano)fibers. The main one involves the direct mixing of magnetic nanoparticles with the polymer solution followed by electrospinning. For obtaining homogeneous nanoparticle dispersions in polymer-containing solutions, mechanical stirring and ultrasonication are usually employed [67, 107, 109]. However, due to their high surface areas and the development of strong magnetic interactions, the magnetic nanoparticles exhibit a high tendency for agglomeration. Consequently, inhomogeneous polymer solutions are generated and solution stability issues that disfavor the electrospinning process rise. As a result the nanoparticles may not be evenly distributed within the produced electrospun nanofibers, whereas such aggregation phenomena may significantly affect the magnetic properties of the final products. For all the above reasons, appropriate stabilizing agents such as surfactant molecules or polymers are introduced together with the magnetic nanoparticles in the polymer solution prior to the electrospinning process [56, 78, 79, 86, 104].

In another approach, a precursor solution containing the polymer and the salt precursor (iron salt precursor) is electrospun first, followed by the immersion of the resulting electrospun mat inside a solution containing appropriate chemical reagents that react with the metal ions attached onto the electrospun polymer nanofibers resulting to the generation of iron oxide magnetic nanoparticles onto the nanofiber surfaces [69].

Table 1. Literature references on polymer-based, electrospun magnetoactive fibrous nanocomposites

Nanoparticle	Reference number	Nanoparticle	Reference number
Polymethacrylates, polyacrylates		*Natural polymers and biopolymers (chitosan and derivatives, cellulose derivatives, DNA)*	
Fe_3O_4	[54; 55; 56; 57; 58; 59]	Fe_3O_4	[60; 61; 62; 63; 64]
Co	[65]	Fe_2O_3	[66]
Polyurethanes		*Polyacrylonitrile (PAN) and copolymers*	
Fe_2O_3	[67]	Fe_3O_4	[68; 69; 70; 71; 72; 73]
MnZnFe-Ni	[74]	$CoFe_2O_4$	[75; 76]
Polyvinylpyrrolidone (PVP)		Fe/FeO	[77]
Fe_3O_4	[78; 79; 64; 80; 81; 82; 83; 84; 85; 86]	*Polyvinyl alcohol (PVA)*	
$Co_xNi_yFe_2O_4$	[87]	Fe_3O_4	[88; 89; 90; 91; 54; 61]
Fe	[92]	Co_3O_4	[93]
Degradable polyesters		*Others**	
Poly(L-lactide) (PLLA)		Fe_3O_4	[94; 95; 96; 97; 98; 99; 100; 101; 102; 103) (104; 63]
Fe_3O_4	[105; 104]	Ni	[106]
Fe_2O_3	[107; 108; 59]	Fe/FeO	[109]
Poly(L-lactide)-co-poly(glycolide) (PLGA)		* including polyvinylchloride, polyaniline, poly(ethylene terephthalate), polymethylglutamide, polyarylene ether nitrile, polystyrene, polyimide, poly(ethylene oxide), poly(p-phenylenevinylene), polystyrene-*block*-polyisoprene, polyvinylidenedifluoride, etc.	
Fe_3O_4	[110; 111; 112; 113]		
Poly(caprolactone) (PCL)			
Fe_3O_4	[114; 115]		
FePt	[116; 117]		

Post-modification processes involving the anchoring of antibody-modified conductive magnetic nanoparticles onto the surface of bio-functionalized electrospun membranes have been also successfully employed aiming toward the development of novel biosensors [118, 66]. Figure 4 provides a general schematic illustration of the above-mentioned fabrication methodologies.

Besides its use in fabricating polymer-based magnetoactive (nano)fibers, electrospinning was also successfully employed for the preparation of magnetoactive metallic nanofibers. One such example is described by H. Wu et al. who prepared electrospun ferromagnetic Fe, Co and Ni transition metal nanofibers by combining electrospinning and sol-gel processes. Polymer solutions containing metal ion precursors were electrospun, followed by calcination at high temperatures and subsequent reduction in a hydrogen atmosphere [119]. In line with the above-mentioned approach, E. Santala and co-workers combined electrospinning with atomic layer deposition to fabricate composite nanofibers consisting of a magnetic core ($NiFe_2O_4$, $CoFe_2O_4$, Fe_2O_3) and a non-magnetic TiO_2 photocatalytic sheath [120].

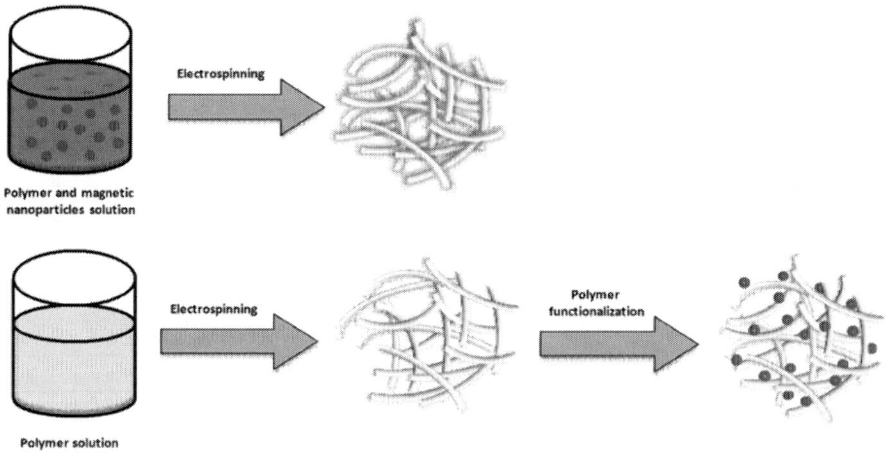

Figure 4. Fabrication strategies employed for generating electrospun magnetoactive polymer-based fibrous nanocomposites. (a) 1-step fabrication process involving the electrospinning of a solution or dispersion containing the polymer and the magnetic nanoparticles. (b) 2-step fabrication process, involving the electrospinning of a polymer-containing solution followed by the post-functionalization of the resulting electrospun fibers with magnetic nanoparticles.

BIOMEDICAL APPLICATIONS OF ELECTROSPUN MAGNETOACTIVE FIBROUS NANOCOMPOSITES

Tissue Engineering

Tissue engineering has emerged in the early 1990s as an excellent approach for the repair/regeneration of damaged or failed tissues, with the potential to circumvent all the limitations of organ transplantation and synthetic tissue replacements. Generally it involves the isolation of healthy cells from a patient, followed by their expansion *in vitro*. This approach is usually achieved by implantation of a three dimensional (3D) biodegradable scaffold, which mimics the natural extracellular matrix (ECM) both, in terms of chemical composition and physical structure, providing structural support and signals, thus promoting cell growth and tissue regeneration [121, 122, 39].

A matrix or culture is required to facilitate cell growth. *In-vivo*, cells are arranged in the ECM compromising a 3D complex structural entity, surrounding and supporting cells. The ECM is necessary for certain cells to carry out their functions. In fact, different parts of the body have many different types of ECMs [29]. In native tissue, the ECM is mainly composed of two classes of macromolecules: polysaccharides which are usually covalently linked to a protein in the form of proteoglycans and fibrous proteins including collagen, elastin, fibronectin and laminin which have both structural and adhesive functions. The proteoglycan molecules form highly hydrated gels, in which the assembled fibrous proteins are embedded and interact with cells through mechanical as well as chemical signals.

During the last decade, electrospun nanofibers have attracted considerable attention in the field of biomaterials and specifically in tissue engineering applications because of their high surface-to-volume ratio, the broadness in regards to materials' chemical composition, the possibility of fiber biofunctionalization, enabling the incorporation of substances that enhance the efficacy of cell adhesion, migration, proliferation and differentiated function, and finally the porosity of the electrospun mats assisting in nutrient transports [123]. Irrespective of their fabrication method, nanofibers have been used as scaffolds in musculoskeletal and connective tissue engineering (including bone and skeletal muscles), skin, vascular and neural tissue engineering [9].

Basic requirements for an effective polymer nanofibrous scaffold in the regeneration of a tissue include a 3D structure of high surface area, high porosity and proper pore size. Additionally, the biodegradability profile is an essential parameter in materials destined for use in tissue engineering applications. The material should not evoke a sustained inflammatory or toxic response upon implantation in the body, and should have an acceptable shelf life. The degradation time of the material should match to the time required for the healing or regeneration process and the degradation products should be non-toxic, and able to get metabolized and cleared from the body. Moreover, the material should have appropriate permeability and mechanical properties for the indicated application [124, 125, 126, 8, 22].

The structural features of tissue engineering scaffolds affect cell response and must be engineered to support cell adhesion, proliferation, migration and differentiation, thereby ensuring functional neotissue formation. During the last years, magnetic-based therapeutic strategies have shown promising results in tissue engineering applications. More precisely, it has been demonstrated that magnetoactive tissue engineering scaffolds can be used in the treatment of bone diseases by promoting proliferation and differentiation of osteoblasts, increasing osteointegration, and accelerating new bone formation [127]. Moreover, a magnetically-modified extracellular matrix protein (ECM) namely laminin has been attached in a gradient manner onto electrospun nanofibrous scaffolds using an externally-applied magnetic field [128]. Since the gradient increase in the concentration of such ECM proteins may significantly enhance the peripheral nerve regeneration, such scaffolds can be very useful in neural tissue engineering.

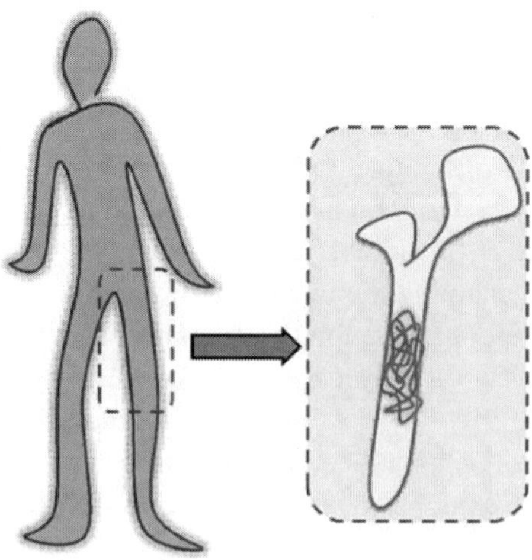

Figure 5. Electrospun nanofibrous membranes employed as scaffolds in bone regeneration processes.

From all the above it becomes obvious that the combination of electrospun nanofibers with magnetic nanoparticles is expected to lead to new materials with superior properties and functions enabling their high efficacy as advanced tissue engineering scaffolds [129, 130, 126].

To the best of our knowledge, very few studies exist describing the fabrication of magnetoactive electrospun fibrous membranes employed as scaffolds for bone regeneration (Figure 5). K. Lai and co-workers have reported on the fabrication of fibrous biocompatible scaffolds consisting of poly(lactic-*co*-glycolic acid) (PLGA) and superparamagnetic Fe_3O_4 nanoparticles and evaluated their effect on different bone cells in the absence of an external magnetic field [66]. The magnetite-containing nanocomposite scaffolds exhibited enhanced osteoblast cell attachment at an early culture time in comparison with the pristine PLGA fibrous analogues.

Y. Wei et al. have reported on the fabrication of magnetic biodegradable materials by means of the electrospinning technique, based on native polysaccharide chitosan (CS), polyvinyl alcohol (PVA), and iron oxide (Fe_3O_4) magnetic nanoparticles with potential use in bone regeneration. MG-63 cells were cultured on the Fe_3O_4/CS/PVA nanofibrous membranes to evaluate the cell growth dynamics. The obtained results demonstrated that cell adhesion and proliferation increased in the presence of the magnetic nanoparticles [61].

The investigation of cell proliferation, differentiation and ECM secretion of osteoblast cells was carried out in the presence of magnetoactive, electrospun nanofibrous composite mats under a static magnetic field [108]. The scaffolds were composed of poly(D,L-lactide) (PLA), hydroxypatite nanoparticles (nHA) and *meso*-2,3-dimercaptosuccinic acid (DMSA)-coated γ-Fe_2O_3 nanoparticles. Osteoblast cells displayed a significantly higher proliferation rate in the presence of the embedded magnetic nanoparticles. Moreover, it has been demonstrated that the static applied magnetic field (0.9-1.0 mT) favored cell proliferation and differentiation and that it might have led to the enhancement of new ECM secretion of the osteoblast cells.

Drug Delivery

Polymer-based materials including macromolecular aggregates such as micelles [131, 132, 61], vesicles [133, 134, 135], capsules [136, 137, 138], networks [139, 140, 141] and fibrous membranes [142, 143] combined with pharmaceutical compounds have been extensively investigated in the biomedical area as drug delivery systems. The ability of such materials to act as controlled drug delivery systems with optimal pharmacokinetic properties is a key requirement in order to avoid the so called “burst release” [144], as well as the generation of undesirable side-effects, due to the uncontrolled drug release process.

During the last few years, there has been a considerable attention on the exploitation of electrospun fibrous materials as drug delivery systems [145, 146, 147, 148, 149, 150, 32]. The unique characteristics of polymer-based (nano)fibrous materials fabricated by means of the electrospinning technique including a very large surface area to volume ratio, the potential to incorporate active chemistry, their high permeability and low basis weight, allow for the biofunctionalization of their surfaces that can be carried out by either chemical or physical modification, with bioactive substances and cell-recognizable ligands [151]. Furthermore, their inherent high surface to volume ratio is advantageous in regards to the drug loading capability and mass transfer properties.

Of great importance in the design and fabrication of new nanosystems to be used in medical treatment are stimuli-responsive polymer-based biomaterials, exhibiting sensitivity to common triggers, including changes in pH, temperature and magnetic or electrical field. These "smart" nanostructures have demonstrated great potential in the biomedical field as controlled drug delivery systems [152, 153, 154].

The incorporation of magnetic nanoparticles within a (nano)fibrous drug delivery platform may be proved to be a very promising therapeutic tool in the future, since it may lead to the development of novel and highly effective multifunctional magnetically-triggered drug delivery systems. Moreover, the presence of the magnetic nanoparticles allows for the future exploitation of such materials in combined therapeutic modalities including hyperthermia treatment [155] and in magnetothermally triggered drug delivery [156].

Even though there is a significant number of studies reported so far on the fabrication of electrospun, polymer-based fibrous (nano)composites comprised of polymeric materials and magnetic nanoparticles, very few deal with the investigation of such systems in drug delivery processes. To the best of our knowledge only four reports so far have been focused on the aforementioned application of such polymer-based nanocomposites.

L. Wang and co-workers, have reported on the fabrication of electrospun nanofibrous composites based on two cellulose derivatives and Fe_3O_4 magnetite nanoparticles [157]. In those, either indomethacin (IDN) or aspirin have been incorporated as model pharmaceutical compounds and their release profile has been investigated in artificial intestinal juice. Based on their studies they have claimed that the presence of magnetic nanoparticles does not influence the drug release profiles of the nanofibrous devices. Moreover, Tan et al. [59] described the fabrication of superparamagnetic polymer nanofibers consisting of poly(hydroxyethyl methacrylate), PHEMA, poly-L-lactide (PLLA) and magnetite nanoparticles by electrospinning and investigated their efficiency in drug delivery *via* the incorporation of a fluorescent enzyme namely albumin with dog fluorescein isothiocyanate (ADFI). Monitoring of the drug release properties of the fluorescent model pharmaceutical compounds has been carried out by means of UV-vis spectrophotometry and fluorescence emission spectroscopy.

In another example tetracycline hydrochloride (TCH)-containing electrospun fibrous nanocomposites comprised of PLLA, PCL and Fe_3O_4 nanoparticles have been studied as drug delivery systems [98]. More precisely, the drug release profiles from the drug-loaded membranes containing or not magnetite nanoparticles were determined spectrophotometrically in phosphate buffer solutions. The nanocomposite fibrous mats exhibited a sustained drug release performance. Moreover it has been demonstrated that the drug release rate was influenced by the presence of magnetite nanoparticles, i.e., the release from the nanocomposite fibrous mats was faster in comparison to the drug release from the pristine analogues.

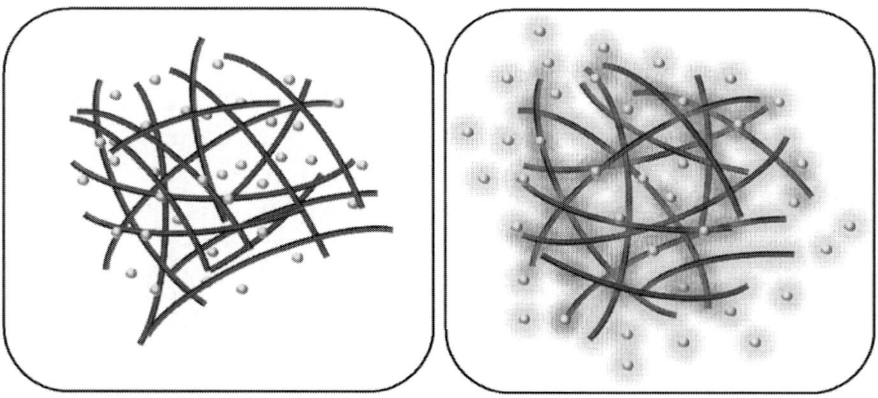

Figure 6. Schematic presentation of the drug release process from drug-loaded magnetoactive electrospun membranes when immersed in an aqueous medium.

Finally, in a very recent work by our group [104] we have described the fabrication of electrospun magnetoactive polyethylene oxide/poly(L-lactide) (PEO/PLLA) membranes with embedded oleic acid-coated magnetite nanoparticles (OA.Fe_3O_4). Those systems exhibited superparamagnetic behavior at room temperature. The membranes were loaded with a widely used analgesic namely acetaminophen (paracetamol) and its release profile was investigated in a buffer solution and in complete culture media (Figure 6). From this investigation in has been demonstrated that the membrane magnetic content as well as the protein content of the release medium are highly influencing factors in the drug release rate. Moreover, the magnetoactive PEO/PLLA/OA.Fe_3O_4 membranes have been also successfully employed for hyperthermia remote induction through AC magnetic field, demonstrating the potential use of such materials in hyperthermia cancer treatment and magnetically-triggered drug delivery therapies.

Hyperthermia

The use of magnetic nanoparticles in the hyperthermia treatment of malignant tumors has been recognized as a highly promising therapeutic modality in cancer treatment. In magnetic hyperthermia, an alternating magnetic field is applied externally at selected sites of the body. The magnetic field is not absorbed by living tissues and it can be applied to deep regions in the living body [158]. Magnetic nanoparticles that are distributed at the targeted tumor site interact with the magnetic field undergoing two different relaxation mechanisms that result to a temperature increase above the physiological body temperature within the range of 41-45 ^{o}C [60]. More precisely, the Néel relaxation mechanism involves rapid directional changes of the magnetic moments relative to crystal lattice whereas in the Brownian relaxation mechanism heat generation is caused by friction phenomena occurring *via* the rotation of the nanoparticles in the medium in which they are dispersed [159].

Within a temperature rise in the range of 41-47 ^{o}C, tumor cells are permanently damaged whereas the healthy cells survive [160]. Despite the selective damage of cancerous cells due to heating phenomena induced by a magnetic field, magnetic hyperthermia renders cancerous cells more vulnerable to other therapeutic modalities including chemotherapy (Figure 7), and radiotherapy [161, 162, 163] resulting to the enhancement of their therapeutic efficacy.

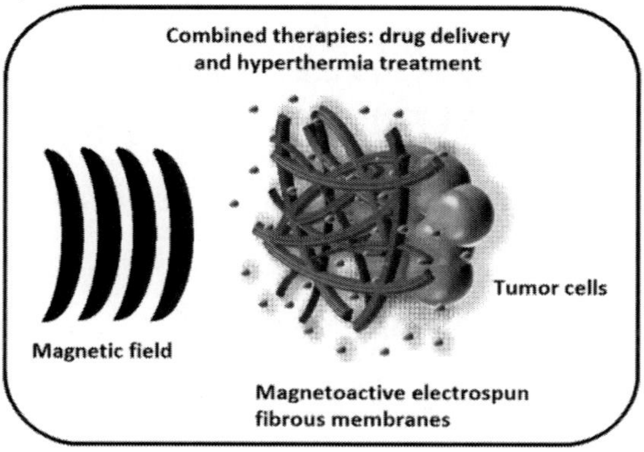

Figure 7. Applicability of magnetoactive electrospun fibrous membranes in combined therapeutic modalities: Drug delivery and hyperthermia cancer treatment.

Numerous literature examples exist, dealing with the magnetic heating properties of different types of magnetic nanoparticles including iron oxides (Fe_3O_4 and Fe_2O_3) [164], $CoFe_2O_4$ and $MnFe_2O_4$ [165] etc. However, the number of reported studies involving the investigation of the applicability of electrospun fibers with embedded magnetic nanoparticles in hyperthermia treatment is very limited.

In a recent study, electrospun biodegradable chitosan nanofibers with embedded magnetite nanoparticles were explored in regards to their heating capability when exposed to an alternating magnetic field [60]. It was found that the heat produced by the magnetoactive electrospun mats *via* the interaction of the embedded magnetite nanoparticles with the applied alternating magnetic field resulted to a temperature increase in the medium to 45 ^{o}C, thus demonstrating the potency of these materials in endoscopic treatment of malignant tumors. Moreover, by performing *in vitro* studies, it has been shown that the generated heat causing the aforementioned temperature rise resulted to a significant reduction in the proliferation and growth rate of malignant cells.

In another example, Fe_2O_3 nanoparticles have been incorporated within polyurethane nanofibers by electrospinning, and the resulting fibrous nanocomposite mats have been exposed to an alternating magnetic field [67]. The heating effect was more pronounced upon increasing the magnetic content within the polymeric nanofibers whereas it had been also demonstrated that the intensity of the applied magnetic field influences the heating capability of the membranes.

Although the applicability of polymer-based electrospun magnetoactive fibrous nanocomposites in hyperthermia cancer treatment is new and still unexplored, the first studies provide very promising results, pointing toward their successful use in cancer therapy, either as a stand-alone curing modality or in combination with other therapeutic methods.

Biosensing

According to IUPAC *"a biosensor is a self-contained integrated device which is capable of providing specific quantitative or semi-quantitative analytical information using a biological recognition element (biochemical receptor) which is in direct spatial contact with a transducer element. A biosensor should be clearly distinguished from a bioanalytical system which requires additional processing steps, such as reagent addition. Furthermore, a biosensor should be distinguished from a bioprobe which is either disposable after one measurement, i.e., single use, or unable to continuously monitor the analyte concentration"* [166].

Biosensors are classified into electrochemical, optical, acoustic and immunosensors. Electrochemical sensing is based on potensiometric, voltammetric, coulometry and conductivity measurements. Optical biosensors which include fluorescence, infrared, photoionization and fiber optics, are based on measurements of light emitted and absorbed due to concentration changes in the sensors' surroundings [167, 168]. Acoustic biosensors including piezoelectric and ultrasonic sensors measure changes in the vibration frequency of the sensing interface, due to changes in the environmental concentration which are reflected as mass changes. Finally, immunosensors, which can be divided into labeled and non-labeled, use antibodies which act as selective binding components. In the former category, the direct measurement of the physical changes caused by the formation of the antigen-antibody complex (immunocomplex) is feasible, whereas in the case of the labeled immunosensors the incorporation of a detectable label is required.

In recent years there has been a considerable interest by researchers worldwide in the direction of development of novel nanofibrous biosensors characterized by high sensitivity and selectivity for the detection of specific molecules (Figure 8). The ease of production and the low fabrication cost, the possibility of chemical and biofunctionalization, the high porosity, the low resistance to diffusion providing excellent permeability to solvents and the high active surface area make electrospun nanofibers ideal candidates in biosensing applications [169, 170].

Polyamide-based electrospun nanofibrous membranes have been examined as glucose biosensors in a variety of food beverages [170]. More precisely, a carbon electrode has been coated with nylon-6 nanofibers in which glucose oxidase (GOD) was covalently tethered. The unique properties exhibited by these nanofibrous materials such as the high surface to volume ratio, high porosity, permeability and selectivity render them ideal for the development of highly efficient glucose biosensors.

In another example, biocomposite membranes consisting of electrospun polyvinyl alcohol (PVA) fibers and GOD were developed and used in constructing amperometric glucose biosensors [171]. The immobilization of GOD in the PVA membranes was carried out *via* chemical crosslinking of the PVA/GOD nanofibers using glutaraldehyde as the crosslinking agent. For comparison reasons, PVA/GOD biosensors have been also fabricated in the form of casting films. The electrospun fibrous-based biosensors exhibited a shorter response time and higher output current to glucose in the normal and diabetic level.

Nanocomposite fibers based on urease and polyvinylpyrrolidone (PVP) have been prepared by means of the electrospinning technique and further tested in potentiometric urea biosensing [172]. Likewise to the previously described examples, these nanofibrous polymer-enzyme materials demonstrated fast response time and high sensitivity to low analyte concentrations, resulting to their potential use not only as medical diagnostic elements but also as environmental and bio-industrial analytical tools.

The design and fabrication of electrospun nanocomposite fibrous membranes consisting of polymers, bioactive molecules and nanostructured materials such as gold nanoparticles [173, 40], carbon nanotubes [174], magnetic nanoparticles [118, 66] and quantum dots [40] may create new pathways for the further advancement of the biosensing field [175, 176].

To the best of our knowledge, only two studies exist so far on the use of electrospun magnetoactive fibrous nanocomposite membranes in biosensing applications [66, 118]. Specifically, lateral-flow immunobiosensors based on electrospun cellulose nitrate nanofibers functionalized with Escherichia coli antibodies and conductive magnetic (Fe_3O_4) nanoparticles have been developed and tested in the detection of pathogenic Escherichia coli O157:H7 bacteria. The increased surface area and mass transfer rates resulted to an enhancement in the binding effect and response of these new biosensing systems.

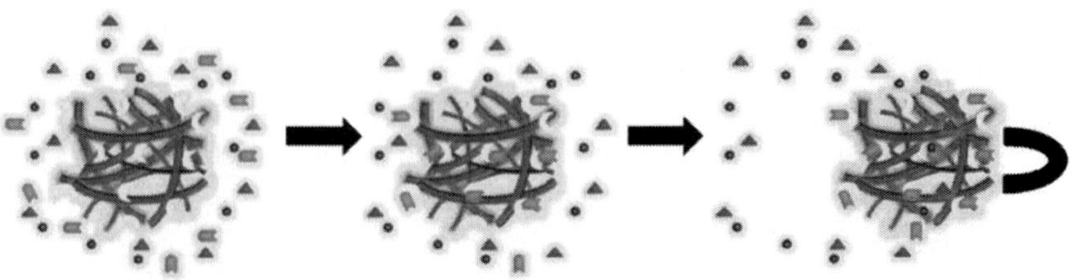

Figure 8. Schematic presentation of a magnetoactive fibrous nanocomposite membrane acting as a selective substrate for specific biological molecule recognition through the development of specific interactions.

Conclusion

During the last decade electrospinning has been proved to be an excellent fabrication technique for the generation of nanofibers destined for use in biomedical applications. Its simplicity, straightforwardness and cost-effectiveness allows for the development of nanofibrous materials comprised of a variety of biocompatible/biodegradable polymers combined with other components that afford additional biomedical functions to these materials. Among others, magnetoactive nanofillers have been incorporated within electrospun polymer nanofibers and further investigated in tissue engineering, drug delivery, cancer hyperthermia treatment and biosensing applications. Although – as discussed in this chapter – the investigation of the applicability of such materials in the biomedical field is still at a premature level and the number of related literature reports is rather limited, the initial outcome demonstrates their high potential in future healthcare applications.

References

[1] Li, Z; Wang, C. One-Dimensional nanostructures electrospinning technique and unique nanofibers. *Springer Briefs in Materials*, 2013.
[2] Joachim, T; Ondarcuhu, C. Drawing a single nanofiber over hundreds of microns. *Europhysics letters*, 1998, Vol. 42, 215-220.
[3] Grzybowski, GM; Whitesides, B. Self-assembly at all scales. *Science*, 2002, Vol. 295, 2418-2421.
[4] Smith, LA; Ma, PX. Nano-fibrous scaffolds for tissue engineering. *Colloids and Surfaces B: Biointerfaces*, 2004, Vol. 39, 125–131.

[5] Ellison, JC; Phatak, A; Giles, DW; et al. Melt blown nanofibers: Fiber diameter distributions and onset of fiber breakup. *Polymer*, 2007, Vol. 48, 3306-3316.

[6] Charlesr, M. Template synthesis of electronically conductive polymer nanostructures. *Accounts Chemicals Research*, 1995, Vol. 28, 61-68.

[7] Teo, WE; Ramakrishna, S. A review on electrospinning design and nanofibre assemblies. *Nanotechnology*, 2006, Vol. 17, R89-R106.

[8] Huang, ZM; Zhang, YZ; Kotaki, M; Ramakrishna, S. A review on polymer nanofibers by electrospinning and their applications in nanocompisites. *Composites Science and Technology*, 2003, Vol. 63, 2223-2253.

[9] Venugopal, J; Zhang, YZ; Ramakrishna, S. Electrospun nanofibers: biomedical applications. *Part N: journal of Nanoengineering and Nanosystems*, 2005, Vol. 218, 35-45.

[10] Reneker, DH; Chun, I. Nanometer diameter fibers of polymer, produced by electrospinning. *Nanotechnology*, 1996, Vol. 7, 216-223.

[11] Venugopal, J; Zhang, YZ; Ramakrishna, S. Electrospun nanofibers: biomedical applications. Part N: *Journal of Nanoengineering and Nanosystems*, 2005, Vol. 218, 35-45.

[12] Yang, J; Zhan, S; Wang, N; Wang, X; Li, Y; Li, Y; Ma, W; Yu, H. A mini review: electrospun hierarchical nanofibers. *Journal of Dispersion Science and Technology*, 2010, Vol. 31, 760-769.

[13] Wang, HS; Fu, GD; Li, XS. Functional polymeric nanofibers from electrospinning. *Recent Patents of Nanotechnology*, 2009, Vol. 3, 21-32.

[14] Li, D; Xia, Y. Electrospinning of nanofibers: Reinventing the wheel. *Advanced Materials*, 2004, Vol. 16, 1151-1170.

[15] Khan, N. Applications of electrospun nanofibers in the biomedical field. *Studies by Undergraduate Researchers at Guelph*, 2012, Vol. 5, 63-73.

[16] Garg, K; Bowlin, GL. Electrospinning jets and nanofibrous structures. *Biomicrofluidics*, 2011, Vol. 5, 1-19.

[17] Frenot, A; Chronakis, IS. Polymer nanofibers assempled by electrospinning. *Current Opinion in Colloid and Interface science*, 2003, Vol. 8, 64-75.

[18] Dong, Z; Kennedy, SJ; Wu, Y. Electrospinning materials for energy-related applications and devices. *Journal of Power Sources*, 2011, Vol. 196, 4886-4904.

[19] Xu, Z; Huang, X; Wan, L. Surface engineering of polymer membranes-Nanofibrous membrane with functionalized surface. *Advanced Topics in Science and Technology*, 2009, 306-328.

[20] Yu, DG; Zhu, LM; White, K; Branford-White, C. Electrospun nanofiber-based drug delivery systems. *Health*, 2009, Vol. 1, 67-75.

[21] Mitchell, SB; Sanders, JE. A unique device for controlled electrospinning. *Journal of bimedical Materials Research*: *Part A*, 2006, 110-120.

[22] Pham, QP; Sharma, U; Mikos, AG. Electrospinning of polymeric nanofibers for tissue engineering applications: A review. *Tissue engineering*, 2006, Vol. 12, 1197-1211.

[23] Tardajos, MG; Nash, M; Rochev, Y; Reinecke, H; Elvira, C; Gallardo, A. Homologous copolymerization route to functional and biocompatible polyvinylpyrrolidone. *Macromolecular Chemistry and Physics*, 2012, Vol. 213, 529−538.

[24] Deitzel, JM; Kleinmeyer, J; Harris, D; Beck Tan, NC. The effect of processing variables on the morphology of electrospun nanofibers and textiles. *Polymer*, 2001, Vol. 42, 261–272.

[25] Yang, J; Zhan, S; Wang, N; Wang, X; Li, Y; Li, Y; Ma, W; Yu, H. A mini review: electrospun hierarchical nanofibers. *Journal of Dispersion Science and Technology*, 2010, Vol. 31, 760-769.

[26] Huang, C; Chen, S; Lai, C; Renker, DH; Qiu, H; Ye, Y; Hou, H. Electrospun polymer nanofibers with small diameters. *Nanotechnology*, 2006, Vol. 17, 1558-1563.

[27] Subbiah, T; Bhat, GS; Tock, RW; Parameswaran, S; Ramkumar, SS. Electrospinning of nanofibers. *Journal of Applied Polymer Science*, 2005, Vol. 96, 557-569.

[28] Sill, TJ; von Recum, AH. Electrospinning: Application in dtyg delivery and tissue engineering. *Biomaterials*, 2008, Vol. 29, 1989-2006.

[29] Greiner, A; Wendorff, JH. Functional self-assembled nanofibers by electrospinning. *Advanced Polymer Science*, 2008, Vol. 219, 107-171.

[30] Baumgarten, P. Electrostatic spinning of acrylic microfibers. *Journal of Colloid Interface Science*, 1971, Vol. 36, 71-79.

[31] Hayati, I; Bailey, AI; Tadros, TF. Investigations into the mechanisms of electrohydrodynamic spraying of liquids: I. Effect of electric field and the environment on pendant drops and factors affecting the formation of stable jets and atomization. *Journal of Colloid Interface Science*, 1987, Vol. 117, 205-221.

[32] Pillay, V; Dott, C; Choonara, YE; Tyagi, C; Tomar, L; Kumar, P; du Toit, LC; Ndesendo, VMK. A review of the effect of processing variables on the fabrication of electrospun nanofibers for drug delivery applications. *Journal of Nanomaterials*, 2013, Vol. 2013, 1-22.

[33] Hohman, MM; Shin, M; Rutledge, G; Brenner, MP. Electrospinning and electrically forced jets. II. Applications. *Physics of Fluids*, 2001, Vol. 13, 2221-2236.

[34] Bhardwaj, N; Kundu, SC. Electrospinning: Afascinating fiber fabrication technique. *Biotechnology Advances*, 2010, 325-347.

[35] Gupta, P; Elkins, C; Long, TE; Wilkes, GL. Electrospinning of linear homopolymers of poly(methyl methacrylate): exploring relationships between fiber formation, viscosity, molecular weight and concentration in a good solvent. *Polymer*, 2005, Vol. 46, 4799–4810.

[36] Pradny, M; Martinova, L; Michalek, J; Fenclova, T; Krumbholcova, E. Electrospinning of the hydrophilic poly (2-hydroxyethyl methacrylate) and its copolymers with 2-ethoxyethyl methacrylate. *Central European Journal of Chemistry*, 2007, Vol. 5, 779-792.

[37] Wannatong, L; Sirivat, A; Supaphol, P. Effects of solvents on electrospun polymeric fibers: preliminary study on polystyrene. *Polymer International*, 2004, Vol. 53, 1851-1859.

[38] Liu, J; Kumar, S. Microscopic polymer cups by electrospinning. *Polymer*, 2005, Vol. 46, 3211-3214.

[39] Supaphol, P; Suwantong, O; Sangsanoh, P; Srinivasan, S; Jayakumar, R; Nair, SV. Electrospinning of biocompatible polymers and their potentials in biomedical applications. *Advanced Polymer Science*, 2012, Vol. 246, 213-240.

[40] Zhang, X; Guo, Q; Cui, D. Recent advances in nanotechnology applied to biosensors. *Sensors*, 2009, Vol. 9, 1033-1053.

[41] Jaeger, R; Bergshoeft, MM; Martin i Batlle, C; Schlinherr, H; Vancso, GJ. Electrospinning of ultra-thin polymer fibers. *Macromolecular Symposia*, 1998, Vol. 127, 141-150.

[42] Megelski, S; Stephens, JS; Bruce Chase, D; Rabolt, JF. Micro- and nanostructured surface morphology on electrospun polymer fibers. *Macromolecules*, 2002, Vol. 35, 8456-8466.

[43] Yuan, X; Zhang, Y; Dong, C; Sheng, J. Morphology of ultrafine polysulfone fibers prepared by electrospinning. *Polymer International*, 2004, Vol. 53, 1704-1710.

[44] Liu, HQ; Hsieh, YL. Ultrafine fibrous cellulose membranes from electrospinning of cellulose acetate. Polymer Science, *B-Polymer Physics*, 2002, Vol. 40, 2119-2129.

[45] Liu, F; Urban, MW. Recent advances and challenges in designing stimuly-responsive polymers. *Progress in Polymer Science*, 2010, Vol. 35, 3-23.

[46] Papaphilippou, P; Christodoulou, M; Marinica, OM; Taculescu, A; Vekas, L; Chrissafis, K; Krasia-Christoforou, T. Multiresponsive polymer conetworks capable of responding to changes in pH, temperature, and magnetic field: synthesis, characterization, and evaluation of their ability for controlled uptake and release of solutes. *Applied Materials and Interfaces*, 2012, Vol. 4, 2139-2147.

[47] Felix, A Plamper; Schmalz, A; Ballauff, M; Muller, AHE. Tuning the thermoresponsiveness of weak polyelectrolytes by pH and light: lower and upper critical-solution temperature of poly(N,N-dimethylaminoethyl methacrylate). *Journal of American Chemistry Society*, 2007, Vol. 129, 14538-14539.

[48] Laloyaux, X; Mathy, B; Nysten, B; Jonas, AM. Surface and bulk collapse transitions of thermoresponsive polymer. *Langmuir*, 2009, Vol. 26, 838-847.

[49] Wu, Y; Jia, W; An, Q; Liu, Y; Chen, J. Multiaction antibacterial nanofibrous membranes fabricated by electrospinning: an excellent system for antibacterial applications. *Nanotechnology*, 2009, Vol. 20, 1-8.

[50] Hoare, T; Santamaria, J; Goya, GF; Irusta, S; Lin, D; Lau, S; Padera, R; Langer, R; Kohane, DS. A magnetically triggered composite membrane for on-demand drug delivery. *Nanoletters*, 2009, Vol. 9, 3651-3657.

[51] Roy, D; Camber, JN; Sumerlin, BS. Future perspectives and recent advances in stimuly-responsive meterials. *Progress in Polymer Science*, 2010, Vol. 35, 278-301.

[52] Jeong, B; Gutowska, A. Lessons from nature: stimuli-responsive polymers and their biomedical applications. *Trends in Biotechnology*, 2002, Vol. 20, 305-311.

[53] Sridhar, R; Sundarrajan, S; Reddy Venugopal, J; Ravichandran, R; Ramakrishna, S. Electrospun inorganic and polymer composite nanofibers for biomedical applications. *Journal of Biomaterials Science, Polymer Edition*, 2013, Vol. 24, 365-385.

[54] Liu, F; Ni, QQ; Murakami, Y. Preparation of magnetic polyvinyl alcohol composite nanofibers with homogenously dispersed nanoparticles and high water resistance. *Textile Research Journal*, 2013, Vol. 83, 510-518.

[55] Qianli, M; Wensheng, Y; Xiangting, D. Electrospinning fabrication and properties of Fe3O4/Eu(BA)(3)phen/PMMA magnetic-photoluminescentbifunctional composite nanoribbons. *Optical Materials*, 2013, Vol. 35, 526-530.

[56] Savva, I; Krekos, G; Taculescu, A; Marinica, O; Vekas, L; Krasia-Christoforou, T. Fabrication and characterization of magnetoresponsive electrospun membranes based on methacrylic random copolymers and magnetite nanoparticles. *Journal of Nanomaterials*, 2012, Vol. 2012, 1-9.

[57] Medina-Castillo, AL; Fernandez-Sanchez, JF; Fernandez-Gutierrez, A. One-step fabrication of multifunctional core-shell fibers by co-electrospinning. *Advanced Functional Materials*, 2011, Vol. 21, 3488-3495.

[58] Pavan Kumar, VS; Jagadeesh Babu, V; Raghuraman, GK; Dhamodharan, R; Natarajan, TS. Giant magnetoresistance of Fe3O4-polymethylmethacrylate nanocomposite aligned fibers via electrospinning. *Journal of Applied Physics*, 2007, Vol. 101, 114317 (4).

[59] Tan, ST; Wendorff, JH; Pietzonka, C; Hong Jia, Z; Qing Wang, G. Biocompatible and biodegradable polymer nanofibers displaying superparamagnetic properties. *A European Journal of Chemical Physics and Physical Chemistry*, 2005, Vol. 6, 1461-1465.

[60] Lin, TC; Lin, FH; Lin, JC. In vitro feasibility study of the use of a magnetic electrospun chitosan nanofiber composite for hyperthermia treatment of tumor cells - hyperthermia. *ACTA Biomaterialia*, 2012, Vol. 8, 2704-2711.

[61] Wei, Y; Zhang, X; Song, Y; Han, B; Hu, X; Wang, X; Lin, Y; Deng, X. Magnetic biodegradable Fe3O4/CS/PVA nanofibrous membranes for bone regeneration. *Biomedical Materials*, 2011, Vol. 6, 1-15.

[62] Navarathne, D; Ner, Y; Jain, M. Fabrication of DNA-magnetite hybrid nanofibers for water detoxification. *Materials Letters*, 2011, Vol. 65, 219-221.

[63] Mincheva, R; Stoilova, O; Penchev, H; Ruskov, T; Spirov, I; Manolova, N; Rashkov, I. Synthesis of polymer-stabilized magneticnanoparticles and fabrication of nanocomposite fibers thereof using electrospinning. *European Polymer Journal*, 2008, Vol. 44, 615-627.

[64] Bin, W; Pei-Pei, Z; Gareth, WR. A simple route to form magnetic chitosan nanoparticles from coaxial-electrospun composite nanofibers. *Journal of Materials Science*, 2013, Vol. 48, 3991-3998.

[65] Kriha, O; Becker, M; Lehmann, M. Connection of hippocampal neurons by magnetically controlled movement of short electrospun polymer fibers - A route to magnetic micromanipulators. *Advanced Materials*, 2007, Vol. 19, 2483.

[66] Luo, Y; Nartker, S; Wiederoder, M; Miller, H; Hochhalter, D; Drzal, LT; Alocilja, EC. Novel biosensor based on electrospun nanofiber and magnetic nanoparticles for the detection of E. coli O157:H7. *IEEE Transactions on Nanotechnology*, 2012, Vol. 11, 676-681.

[67] Park, CH; Kang, SJ; Tijing, LD; Pant, HR; Kim, CS. Inductive heating of electrospun Fe2O3/polyurethane composite mat under high-frequency magnetic field - hyperthermia. *Ceramics International*, 2013, Vol. 39, 9785-9790.

[68] Wang, X; He, T; Li, D. Electromagnetic properties of hollow PAN/Fe3O4 composite nanofibres via coaxial electrospinning. *International Journal of Materials and Product Technology*, 2013, Vol. 46, 95-105.

[69] Yang, F; Zheng, W; Huang, H. Decoration of electrospun nanofibers with magnetic nanoparticles via electrospinning and sol-gel process. *Chemical research in Chinese Universities*, 2010, Vol. 26.

[70] Wang, B; sun, Y; Wang, H. Preparation and properties of electrospun PAN/Fe3O4 magnetic nanofibers. *Journal of Applied Polymer Science*, 2010, Vol. 115, 1781-1786.

[71] Ye, XY; Liu, ZM; Wang, ZG. Preparation and characterization of magneticnanofibrous composite membranes with catalytic activity. *Materials Letters*, 2009, Vol. 63, 1810-1813.

[72] Zhang, D; Karki, AB; Rutman, D. Electrospun polyacrylonitrile-nanocomposite fibers reinforced with Fe3O4 nanoparticles: Fabrication and property analysis. *Polymer*, 2009, Vol. 50, 4189-4198.

[73] Guo, J; Ye, X; Liu, W. Preparation and characterization of poly(acrylonitrile-co-acrylic acid) nanofibrous composites with Fe3O4 magnetic nanoparticles. *Materials letters*, 2009, Vol. 63, 1326-1328.

[74] Gupta, P; smatulu, R Claus. Superparamagnetic flexible substrates based on submicron electrospun Estane (R) fibers containing MnZnFe-Ni nanoparticles. *Journal of Applied Polymer Science*, 2006, Vol. 100, 4935-4942.

[75] Jin, X; Ni, QQ; Fu, Y. Electrospun nanocomposite polyacrylonitrile fibers containing carbon nanotubes and cobalt ferrite. *Polymer Composites*, 2012, Vol. 33, 317-323.

[76] Chen, IH; Wang, CC; Chen, CY. Fabrication and characterization of magnetic cobalt ferrite/polyacrylonitrile and cobalt ferrite/carbon nanofibers by electrospinning. *Carbon*, 2010, Vol. 48, 604-611.

[77] Zhu, J; Wei, S; Rutman, D. agneticpolyacrylonitrile-Fe@FeO nanocomposite fibers - Electrospinning, stabilization and carbonization. *Polymer*, 2011, Vol. 52, 2947-2955.

[78] Zhang, P; Wang, B; Williams, GR; Branford-White, C; Quan, J; Nie, H; Zhu, L. Self-assembled core-shell Fe3O4@SiO2 nanoparticles from electrospun fibers. *Materials Research Bulletin*, 2013, Vol. 48, 3058-3064.

[79] Guangqing, G; Liyan, W; Xiangting, D. Electrospun Fe3O4/PVP //Tb(BA)(3)phen/PVP magnetic-photoluminescent bifunctional bistrand aligned composite nanofibers bundles. *Journal of Materials Science*, 2013, Vol. 48, 5140-5147.

[80] Ma, Q; Wang, J; Dong, X; Yu, W; Liu, G. Electrospinning fabrication of high-performance magnetic@photoluminescentbifunctional coaxial nanocables. *Chemical Engineering Journal*, 2013, Vol. 222, 16-22.

[81] Gai, G; Wang, L; Dong, X; Zheng, C; Yu, W; Wang, J; Xiao, X. Electrospinning preparation and properties of magnetic-photoluminescent bifunctionalbistrand-aligned composite nanofibers bundles. *Journal of Nanoparticle Research*, 2013, Vol. 15, p. 1539 (9).

[82] Ma, Q; Yu, W; Dong, X. Electrospinning preparation and properties of Fe3O4/Eu(BA)(3)phen/PVP magnetic-photoluminescentbifunctional composite nanofibers. *Journal of Nanoparticle Research*, 2012, Vol. 14.

[83] Chen, M; Qu, H; Zhu, J. Magnetic electrospun fluorescent polyvinylpyrrolidonenanocomposite fibers. *Polymer*, 2012, Vol. 53, 4501-4511.

[84] Ma, Q; Wang, J; Dong, X. Electrospinning preparation and properties of magnetic-photoluminescent bifunctional coaxial nanofibers. *Journal of Materials Chemistry*, 2012, Vol. 22, 14438-14442.

[85] Lin, CR; Tsai, TC; Chung, M; Lu, SZ. Synthesis and characterization of magnetic nanoparticles embedded in polyvinyl pyrrolidonenanofiber film by electrospinning method. *Journal of Applied Physics*, 2009, Vol. 105, p. 07B509 (3).

[86] Savva, I; Constantinou, D; Marinica, O; Vasile, E; Vekas, L; Krasia-Christoforou, T. Fabrication and characterization of superparamagnetic poly(vinyl pyrrolidone)/poly(L-lactide) electrospun membranes. *Journal of Magnetism and Magnetic Materials*, 2014, Vol. 352, 35-40.

[87] Tang, Y; Wang, X; Zhang, Q. Fabrication and characterization of high solid content Co0.8Ni0.2Fe2O4/PVP composite nanofibers by direct-dispersed electrospinning. *Synthetic Metals*, 2012, Vol. 162, 309-313.

[88] Gao, Q; Takizawa, J; Kimura, M. Hydrophilic non-wovens made of cross-linked fully-hydrolyzed poly(vinyl alcohol) electrospun nanofibers. *Polymer*, 2013, Vol. 54, 120-126.

[89] Wang, S; Wang, C; Zhang, B. Preparation of Fe3O4/PVA nanofibers via combining in-situ composite with electrospinning. Materials Letters, 2010, Vol. 64, 9-11.

[90] Babinec, P; Jirsak, O. Microwave absorbing nonwoven textile from electrospun magnetically responsive nanofibres. *Optoelectronics and Advanced Materials-Rabid Communications*, 2008, Vol. 2, 474-477.

[91] Wang, A; Singh, H; Hatton, Ta. Field-responsive superparamagnetic composite nanofibers by electrospinning. *Polymer*, 2004, Vol. 45, 5505-5514.

[92] Xu, X; Wang, Q; Choi, HC. Encapsulation of iron nanoparticles with PVP nanofibrous membranes to maintain their catalytic activity. *Journal of Membrane Science*, 2010, Vol. 348, 231-237.

[93] Jung, JH; Hyun, YH; Park, SY; Lee, YP. Giant superparamagnetic nanocomposites using ferritin. *Journal of the Korean Physical Society*, 2011, Vol. 58, 797-800.

[94] Liu, J; Shi, J; Jiang, L. Segmented magneticnanofibers for single cell manipulation. *Applied Surface Science*, 2012, Vol. 258, 7530-7535.

[95] Ahn, BW; Kang, TJ. Preparation and characterization of magneticnanofibers with iron oxide nanoparticles and poly(ethylene terephthalate). *Journal of Applied Polymer Science*, 2012, Vol. 125, 1567-1575.

[96] Sung, YK; Ahn, BW; Kang, TJ. Magneticnanofibers with core (Fe3O4 nanoparticle suspension)/sheath (poly ethylene terephthalate) structure fabricated by coaxial electrospinning. *Journal of Magnetism and Magnetic Materials*, 2012, Vol. 324, 916-922.

[97] Meng, F; Zhan, Y; Lei, Y. Electrospun magnetic fibrillar polyarylene ether nitriles nanocomposites reinforced with Fe-phthalocyanine/Fe3O4 hybrid microspheres. *Journal of Applied Polymer Science*, 2012, Vol. 123, 1732-1739.

[98] Chiscan, O; Dumitru, I; Postolache, P. Electrospun PVC/Fe3O4 composite nanofibers for microwave absorption applications. *Materials Letters*, 2012, Vol. 68, 251-254.

[99] Sharma, N; Jaffari, GH; Shah, SI; Pochan, DJ. Orientation-dependent magnetic behavior in aligned nanoparticle arrays constructed by coaxial electrospinning. *Nanotechnology*, 2010, Vol. 21, p. 085707 (6).

[100] Yu, QZ. Polyaniline/Au&Fe3O4@Au sub-microcables fabricated by electrospinning and electroless deposition. *Materials Science and Engineering B-Advanced Functional Sold State Materials*, 2010, Vol. 167, 26-30.

[101] Xin, Y; Huang, ZH; Peng, L; Wang, DJ. Photoelectric performance of poly(p-phenylenevinylene)/Fe3O4 nanofiber array. *Journal of Applied Physics*, 2009, Vol. 105, p. 086106 (3).

[102] Karla, V; Lee, J; Lee, JH. Controlling nanoparticle location via confined assembly in electrospun block copolymer nanofiber. *Small*, 2008, Vol. 4, 2067-2073.

[103] Andrew, JS; Clarke, DR. Enhanced ferroelectric phase content of polyvinylidenedifluoride fibers with the addition of magneticnanoparticles. *Langmuir*, 2008, Vol. 24, 8435-8438.

[104] Savva, I; Odysseos, AD; Evaggelou, L; Marinica, O; Vasile, E; Vekas, L; Sarigiannis, Y; Krasia-Christoforou, T. Synthesis, characterization and evaluation in drug release properties of magnetoactive poly(ethylene oxide)-poly(L-lactide) electrospun membranes. *Biomacromolecules*, 2013, Vol. 14, 4436-4446.

[105] Li, HY; Chang, CM; Hsu, KY. Poly(lactide)-functionalized and Fe3O4 nanoparticle-decorated multiwalled carbon nanotubes for preparation of electrically-conductive and magnetic poly(lactide) films and electrospun nanofibers. *Journal of Material Chemistry*, 2012, Vol. 22, 4855-4860.

[106] Chen, X; Wei, s; Gunesoglu, C. Electrospun magneticFibrillar polystyrene nanocomposites reinforced with nickel nanoparticles. *Macromolecular Chemistry and Physics*, 2010, Vol. 211, 1775-1783.

[107] Meng, J; Xiao, B; Zhang, Y; Liu, J; Xue, H; Lei, J; Kong, H; Huang, Y. Super-paramagnetic responsive nanofibrous scaffolds under static magnetic field enhance osteogenesis for bone repair in vivo. *Scientific Reports*, 2013, Vol. 3, 1-7.

[108] Meng, J; Zhang, Y; Qi, X; Kong, H; Wang, C; Xu, Z; Xie, S; Gu, N; Xu, H. Paramagnetic nanofibrous composite films enhance the osteogenic responses of pre-osteoblast cells. *Nanoscale*, 2010, Vol. 2, 2565-2569.

[109] Zhu, J; wei, S; Chen, X. Electrospun polyimide nanocomposite fibers reinforced with core-shell Fe-FeO nanoparticles. *Journal of Physical Chemistry C*, 2010, Vol. 114, 8844-8850.

[110] Li, P; Liu, C; Song, Y. nfluence of Fe3O4 nanoparticles on the preparation of aligned PLGA electrospun fibers induced by magnetic field. *Journal of Nanomaterials*, 2013.

[111] Hu, H; Jiang, W; Lan, F. Synergic effect of magneticnanoparticles on the electrospun aligned superparamagneticnanofibers as a potential tissue engineering scaffold. *RSC Advances*, 2013, Vol. 3, 879-886.
[112] Lee, WY; Cheng, WY; Yeh, YC. Magnetically directed self-assembly of electrospun superparamagnetic fibrous bundles to form three-dimensional tissues with a highly ordered architecture. *Tissue Engineering Part C-Methods*, 2011, Vol. 17, 651-661.
[113] Lai, K; Jiang, W; Tang, JZ; Wu, Y; He, B; Wang, G; Gu, Z. Superparamagnetic nano-composite scaffolds for promoting bone celproliferation and defect reparation without a magnetic field. *RSC Advances*, 2012, Vol. 2, 13007-13017.
[114] Gong, T; Li, W; Chen, H. Remotely actuated shape memory effect of electrospun composite nanofibers. *ACTA Biomaterialia*, 2012, Vol. 8, 1248-1259.
[115] Kannarkat, JT; Battogtokh, J; Philip, J. Embedding of magneticnanoparticles in polycaprolactonenanofiber scaffolds to facilitate bone healing and regeneration. *Journal of Applied Physics*, 2010, Vol. 107.
[116] Song, T; Zhang, YZ; Zhou, TJ. Fabrication of magnetic composite nanofibers of poly(epsilon-caprolactone) with FePtnanoparticles by coaxial electrospinning. *Journal of Magnetism and Magnetic Materials*, 2006, Vol. 303, E286-E289.
[117] Song, T; Zhang, Y; Zhou, T; Teck Lim, C; Ramakrishna, S; Liu, B. Encapsulation of self-assembled FePt magnetic nanoparticles in PCL nanofibers by coaxial electrospinning. *Chemical Physics Letters*, 2005, Vol. 415, 317-322.
[118] Luo, Y; Nartker, S; Miller, H; Hochhalter, D; Wiederoder, M; Wiederoder, S; Setterington, E; Drzal, LT; Alocilja, EC. Surface functionalization of electrospun nanofibers for detecting E. coli O157:H7 and BVDV cells in a direct-charge transfer biosensor. *Biosensors and Bioelectronics*, 2010, Vol. 26, 1612-1617.
[119] Wu, H; Zhang, R; Liu, X; Lin, D; Pan, W. Electrospinning of Fe, Co and Ni nanofibers: synthesis, assembly and magnetic properties. *Materials Chemistry*, 2007, Vol. 19, 3506-3511.
[120] Santala, E; Kemell, M. The preparation of reusable magnetic and photocatalytic composite nanofibers by electrospinning and atomic layer deposition. *Nanotechnology*, 2009, Vol. 20, p. 035602.

[121] Jagur-Grodzinski, J. Polymers for tissue engineering, medical devices, and regenerative medicine. Concise general review of recent studies. *Polymers for Advanced Technologies*, 2006, Vol. 17, 395-418.

[122] Ma, Z; Kotaki, M; Inai, R; Ramakrishna, S. Potential of nanofiber matrix as tissue engineering scaffolds. *Tissue Engineering*, 2005, Vol. 11, 101-109.

[123] Cui, W; Zhou, Y; Chang, J. Electrospun nanofibrous materials for tissue engineering and drug delivery. *Science and Technology of Advanced Materials*, 2010, Vol. 11, 1-11.

[124] Renker, DH; Fong, H. Polymeric nanofibers. *ACS Division of Polymer Chemistry*, Inc, 2005, p. Chapter 13 and 14.

[125] Schmidt, CE; Leach, JB. Neural tissue engineering: Strategies for repair and regeneration. *Annual Review of Biomedical Engineering*, 2003, Vol. 5, 293–347.

[126] Porter, JR; Ruckh, TT; Popat, KC. Bone tissue engineering: A review in bone biomimetics and drug delivery strategies. *Biotechnology Progress (AlChE)*, 2009, Vol. 25, 1539-1560.

[127] Zeng, XB; Hu, H; Xie, LQ; Lan, F; Jiang, W; Wu, Y; Gu, ZW. *Magnetic responsive hydroxypatite composite scaffolds construction for bone defect reparation*, 2012, 3365-3378.

[128] Valmikinathan, MC; Wang, J; Smiriglio, S; Golwala, NG; Yu, X. Magnetically-induced protein gradients on electrospun nanofibers. *Combinatorial Chemistry & High Throughput Screening*, 2009, Vol. 12, 656-663.

[129] Smith, IO; Liu, XH; Smith, LA; Ma, PX. Nano-structured polymer scaffolds for tissue engineering and regenerative medicine. *Wiley Interdisciplinary Reviews Nanomedicine and Nanobiotechnology*, 2009, Vol. 2, 226-236.

[130] Pankhurst, QA; Connolly, J; Jones, SK; Dobson, J. Applications of magnetic nanoparticles in bimedicine. *Journal of Physics D: Applied Physics*, 2003, Vol. 36, 167-181.

[131] Hua, W; Ren-Xi, Z; Xian-Zheng, Z. Design and development of polymeric micelles with cleavable links for intracellular drug delivery. *Progress in Polymer Science*, 2013, Vol. 38, 503-535.

[132] K Glen S; Kazunori, K. Block copolymer micelles as long-circulating drug vehicles. *Advanced Drug Delivery Reviews*, 2012, Vol. 64, 237-245.

[133] Elizondo, E; Veciana, J; Ventosa, N. Nanostructuring molecular materials as particles and vesicles for drug delivery, using compressed and supercritical fluids. *Nanomedicine*, 2012, Vol. 7, 1391-1408.

[134] Tanner, P; Baumann, P; Enea, R; Onaca, O; Palivan, C; Meier, W. Polymeric vesicles: From drug carriers to nanoreactors and artificial organelles. *Accounts of Chemical Research*, 2011, Vol. 44, 1039-1049.

[135] Brinkhuis, RP; Rutjes, FPJT; van Hest, JCM. Polymeric vesicles in biomedical applications. *Polymer Chemistry*, 2011, Vol. 2, 1449-1462.

[136] De Temmerman, ML; Demeester, J; De Smedt, SC; Rejman, J. Tailoring layer-by-layer capsules for biomedical applications. *Nanomedicine*, 2012, Vol. 7, 771-788.

[137] Delcea, M; Möhwald, H; Skirtach, AG. Stimuli-responsive LbL capsules and nanoshells for drug delivery. *Advanced Drug Delivery Reviews*, 2011, Vol. 63, 730-747.

[138] De Koker, S; Hoogenboom, R; De Geest, BG. Polymeric multilayer capsules in drug delivery. *Angewandte Chemie-International Edition*, 2010, Vol. 49, 6954-6973.

[139] Alvarez-Lorenzo, C; Blanco-Fermandez, B; Puga, AM. Crosslinked ionic polysaccharides for stimuli-sensitive drug delivery. *Advanced Drug Delivery Reviews*, 2013, Vol. 65, 1148-1171.

[140] Yong, Q; Park, K. Environment-sensitive hydrogels for drug delivery. *Advanced Drug Delivery Reviews*, 2012, Vol. 64, 49-60.

[141] Chaoliang, H; Sung Wan, K; Doo Sung, L. In situ gelling stimuli-sensitive block copolymer hydrogels for drug delivery. *Journal of Controlled Release*, 2008, Vol. 127, 189-207.

[142] Kumbar, SG; Nair, SL; Bhattacharyya, S; Laurencin, CT. Polymeric nanofibers as novel carriers for the delivery of therapeutic molecules. *Journal of Nanoscience and Nanotechnology*, 2006, Vol. 6, 2591-2607.

[143] Chakraborty, S; Liao, IC; Adler, A; Leong, KW. Electrohydrodynamics: A facile technique to fabricate drug delivery systems. *Advanced Drug Delivery Reviews*, 2009, Vol. 61, 1043-1054.

[144] Huang, X; Brazel, CS. On the importance and mechanisms of burst release in matrix-controlled drug delivery systems. *Journal of Controlled Release*, 2001, Vol. 73, 121-136.

[145] Ignatova, M; Rashkov, I; Manolova, N. Drug-loaded electrospun materials in wound-dressing applications and in local cancer treatment. *Expert Opinion on Drug Delivery*, Vol. 10, 469-483.

[146] Zamani, M; Prabhakaran, MP; Ramakrishna, S. Advances in drug delivery via electrospun and electrosprayed nanomaterials. *International Journal of Nanomedicine*, 2013, Vol. 8, 2997-3017.

[147] Meinel, AJ; Germershaus, O; Luhmann, T; Merkle, HP; Meinel, L. Electrospun matrices for localized drug delivery: Current technologies and selected biomedical applications. *European Journal of Pharamaceutics and Biopharmaceutics*, 2012, Vol. 81, 1-13.

[148] Prabaharan, M; Jayakumar, R; Nair, SV. Electrospun nanofibrous scaffolds-current status and prospects in drug delivery. *Biomedical Applications of Polymeric Nanofibers, Book Series: Advances in Polymer Science*, 2012, Vol. 246, 241-262.

[149] Ignatious, F; Sun, L; Lee, CP; Baldoni, J. *Electrospun Nanofibers in Oral Drug Delivery*, 2010, Vol. 27, 576-588.

[150] Ashammakhi, N; Wimpenny, I; Nikkola, L; Yang, Y. Electrospinning: Methods and Development of Biodegradable Nanofibres for Drug Release. *Journal of Biomedical Nanotechnology*, 2009, Vol. 5, 1-19.

[151] Beachley, V; Wen, X. Polymer nanofibrous structures: Fabrication, biofunctionalization, and cell interactions. *Progress in Polymer Science*, 2010, Vol. 35, 868-892.

[152] Yoshida, T; Lai, TC; Kwon, GS; Sako, K. pH- and ion-sensitive polymers for drug delivery. *Expert Opinion on Drug Delivery*, 2013, Vol. 10, 1497-1513.

[153] Qing, G; Li, M; Deng, L; Lv, Z; Ding, P; Sun, T. Smart Drug Release Systems Based on Stimuli-Responsive Polymers. *Mini-Reviews in Medicinal Chemistry*, 2013, Vol. 65, 1369-1380.

[154] Chan, A; Orme, RP; Fricker, RA; Roach, P. Remote and local control of stimuli responsive materials for therapeutic applications. *Advanced Drug Delivery Reviews*, 2013, Vol. 65, 497-514.

[155] Huang, C; Soenen, SJ; Rejman, J; Trekker, J; Chengxun, L; Lagae, L; Ceelen, W; Wilhelm, C; Demeester, J; De Smedt, SC. Magnetic electrospun fibers for cancer therapy. *Advanced Functional Materials*, 2012, Vol. 22, 2479-2486.

[156] Brazel, CS. Magnetothermally-responsive Nanomaterials: Combining Magnetic nanostructures and thermally-sensitive polymers for triggered drug release. *Pharmaceutical Research*, 2009, Vol. 26, 644-656.

[157] Wang, L; Wang, M; Tophamc, PD; Huang, Y. Fabrication of magnetic drug-loaded polymeric composite nanofibres and their drug release characteristics. *RSC Advances*, 2012, Vol. 2, 2433–2438.

[158] Gupta, AK; Gupta, M. Synthesis and surface engineering of iron oxide nanoparticles for biomedical applications. *Biomaterials*, 2005, Vol. 2006, 3995–4021.

[159] Kumar a, CSSR; Mohammad, F. Magnetic nanomaterials for hyperthermia-based therapy and controlled drug delivery. *Advanced Drug Delivery Reviews*, 2011, Vol. 63, 789–808.

[160] Kobayashi, T. Cancer hyperthermia using magnetic nanoparticles. *Biotechnology Journal*, 2011, Vol. 6, 1342-1347.

[161] Krasia-Christoforou, T; Georgiou, TK. Polymeric theranostics: using polymer-based systems for simultaneous imaging and therapy. *Journal of Material Chemistry B*, 2013, Vol. 1, 3002-3025.

[162] Issels, RD. Hyperthermia adds to chemotherapy. *European Journal of Cancer*, 2008, Vol. 44, 2546-2554.

[163] Wust, P; Hildebrandt, B; Sreenivasa, G; Rau, B; Gellermann, J; Riess, H; Felix, R; Schlag, PM. Hyperthermia in combined treatment of cancer. *Lancet Oncology*, 2002, Vol. 3, 487-497.

[164] Hilger, I; Kaiser, AW. Iron oxide-based nanostructures for MRI and magnetic hyperthermia. *Nanomedicine*, 2012, Vol. 7, 1443-1459.

[165] Sharifi, I; Shokrollahi, H; Amiri, S. Ferrite-based magnetic nanofluids used in hyperthermia applications. *Journal of Magnetism and Magnetic Materials*, 2012, Vol. 324, 903-915.

[166] Frieder, WS; Wollenberger, U; Warsinke, A; Lisdat, F. Research and development in biosensors. *Analytical Biotechnology*, 2001, Vol. 12, 35-40.

[167] Ramakrishna, S; Lala, NL; Garudadhwaj, H; Ramaseshan, R; Ganesh, VK. Polymer nanofibers for biosensor applications. Molecular Building Blocks for Nanotechnology, *Topics in Applied Physics*, 2007, Vol. 109, 377-392.

[168] Gerard, M; Chaubey, A; Malhotra, BD. Application of conductingpolymers to biosensors. *Biosensors and Bioelectronics*, 2002, Vol. 17, 345-359.

[169] Scampicchio, M; Bulbarello, A; Arecchi, A; Cosio, MS; Benedetti, S; Mannini, S. Electrospun nonwoven nanofibrous membranes for sensors and biosensors. *Electroanalysis*, 2012, Vol. 24, 719-725.

[170] Scampicchio, M; Arecchi, A; Lawrence, NS; Mannino, S. Nylon nanofibrous membrane for mediated glucose biosensing. *Sensors and Actuators B*, 2010, Vol. 145, 394-397.

[171] Ren, G; Xu, X; Liu, Q; Cheng, J; Yuan, X; Wu, L; Wan, Y. Electrospun poly(vinyl alcohol)/glucose oxidase biocomposite membranes for

biosensor applications. *Peactive and Functional Polymers*, 2006, Vol. 66, 1559-1564.

[172] Sawickaa, K; Goumab, P; Simon, S. Electrospun biocomposite nanofibers for urea biosensing. *Sensors and Actuators B*, 2005, Vol. 108, 585-588.

[173] Zhang, T; Wang, W; Zhang, D; Zhang, X; Ma, Y; Zhou, Y; Qi, L. Biotemplated synthesis of gold nanoparticle-bacteria cellulose nanofiber nanocomposites and their application in biosensing. *Advanced Functional Materials*, 2010, Vol. 20, 1152-1160.

[174] Jianrong, C; Yuqing, M; Nongyue, H; Xiaohua, W; Sijiao, L. Nanotechnology and biosensors. *Biotechnology Advances*, 2004, Vol. 22, 505-518.

[175] Hasirci, V; Vrana, E; Zorlutuna, P; Ndreu, A; Yilgor, P; Basmanav, FB; Aydin, E. Nanomaterials: a review of the excisting science and technology, and new approaches. *Journal of Biomaterials Science, Polymer Edition*, 2006, Vol. 17, 1241-1268.

[176] Liu, J; Shi, J; Jiang, L; Zhang, F; Wang, L; Yamamoto, S; Takano, M; Chang, M; Zhang, H; Chen, Y. Segmented magnetic nanofibers for single cell manipulation. *Applied Surface Science*, 2012, Vol. 258, 7530-7535.

In: Magnetic Nanoparticles
Editor: Nora P. Sabbas
ISBN: 978-1-63117-434-6

Chapter 7

MAGNETIC NANOCARRIERS FOR TARGETED DRUG DELIVERY

Maja Leitgeb, Gordana Hojnik Podrepšek and Željko Knez

University of Maribor, Faculty of Chemistry and Chemical Engineering, Laboratory for Separation Processes and Product Design; Smetanova, Maribor, Slovenia

ABSTRACT

Interest to the nanomedicine has increased dramatically during past years. Since nanotechnology has the potential to have a revolutionary impact on cancer diagnosis and therapy, is there a major challenge to produce magnetic nanoparticles (MNPs) suitable for biomedical applications. In this review the main theoretical views on the properties, synthesis, characterization, functionalization, and bioapplication of MNPs in the current context of science are considered. This review examines some of the recent developments in MNPs technology and provides a brief background of their applications and results of *in-vitro* and *in-vivo*, animal and clinical experiments of targeted drug delivery. Finally, some of the recent biological, medical and scientific applications of MNPs are briefly reviewed, and some future trends and perspectives in these research areas will be outlined.

1. INTRODUCTION

Nanotechnology is an interdisciplinary field of technological developments on the nanometer scale (10^{-9} m) offering comprehensive applications even to biomedicine. Due to their small size and surface area characteristics, nanoparticles exhibit unique electronic, optical, and magnetic properties that can be exploited for drug delivery. The importance of targeted drug delivery and targeted drug therapy is to transport a drug directly to the center of the disease under various conditions and thereby treat it deliberately, with no effects on the body [1]. Nanoparticle technology may improve the therapeutic index of drugs by enhancing their efficacy and increasing their tolerability in the body. Hence, biomedical applications of nanoparticles generally can be used for diagnostics and therapeutics with the promising technologies and potential uses in biosensing, imaging [2, 3], implantable materials, tissue regeneration and drug delivery especially for cancer treatment [4]. For all these biomedical and bioengineering applications are magnetically guided drug using MNPs a promising approach. Usually, iron oxide nanoparticles, like magnetite and maghemite are used as the artificial sources of magnetic fields [4-16]. Ferrimagnetic magnetite (Fe_3O_4) and maghemite (γ-Fe_2O_3) are the materials mostly suitable for *in-vivo* applications of the MNPs because others such as cobalt and nickel are toxic [17]. MNPs are because of their unique properties in size and structure and their huge surface-to-volume ratio perfect candidates for the targeted drug delivery [18]. Drug targeting has emerged as one of the modern technologies for drug delivery. Drug delivery systems can be specifically designed to control the release rate and/or amount of therapeutic agents that will allow desired effects on target sites. There are a few classes of biocompatible materials used as carrier matrix for drug delivery systems, which include also MNPs. In combination with the external magnetic field MNPs allow delivering of the particles to the desired target area and fixing them at the local site while the medication is released and acts locally [19]. Advantageously, use of a biological targeting agent in the MNPs drug delivery system serves to increase the targeting efficiency for the pharmaceutical agent while reducing the requisite quantity of magnetic nanoparticles, which may be toxic at elevated concentrations [20]. The MNPs applications have drastically increased in recent years. In clinical area of human medicine, these particles are being used as delivery systems for drugs [21], genes [22] and radionuclides [19], also in medical diagnostics by fluorescent imaging [12] or magnetic resonance imaging (MRI) [23]. A wide variety of therapeutic pharmaceutical agents may be employed, either alone or in combination, including but not limited to anticancer drugs and autoimmune/anti-inflammation drugs, disease modifying anti-rheumatic drugs, corticosteroids, etc. [20].

This review examines some of the recent developments in MNPs technology and provides a brief background of their applications and results of *in-vitro* and *in-vivo*, animal and clinical experiments of targeted drug delivery.

2. Magnetic Nanoparticles

2.1. Magnetic Properties

MNPs are increasingly being considered for a number of biomedical applications due to their ultra-fine size, functionality, stability, biocompatibility and superparamagnetic properties [23]. They have been utilized for specific biological functions as nanocarriers for drugs [24], contrast imaging agents in magnetic resonance imaging (MRI) [2, 25], in local hyperthermia [16, 26], cell labeling and sorting [27], magnetic targeting [8, 21], and immunoassays [28]. Typically, a MNPs consists of a magnetically active core (usually magnetite or maghemite) coated with a stabilizing agent as biocompatible polymer to which targeting ligands are embedded in the shell structure or chemically bonded to its surface [29]. Many bioactive substances such as enzymes, proteins, antibodies, and anticancer agents have already been bound to it [8, 30-32]. Recently, a new method for the direct binding of proteins such as bovine serum albumin (BSA) via carbodiimide activation was reported [23]. This method is notable due to its simplicity and high efficiency.
Usually, the size of the particles refers to the total diameter of the particles, including the iron core and the coating. MNPs core diameter should be of less than 20 nm and overall hydrodynamic diameter of less than 50 nm, which is comparable to the nuclear pore size (~50 nm) and much smaller than cell (normally 10-100 nm) [33]. Figure 1 shows the particle size distribution of maghemite in aqueous solution, which exhibits the formation of a spherical structure with a diameter of 22,78 nm. Nanoparticles have a tendency to aggregate, as a function of magnetic energy, which may further effect on increase in size and consequently effect on the drug delivery process [19]. The nanoparticles must be small so that they can be superparamagnetic in order to avoid agglomeration outside of the magnetic field and to remain in circulation without being removed by the body's natural filters such as the liver or the immune system [1]. It was also found out that the shape of the nanoparticles

could affect their biodistribution, clearance, and biocompatibility *in-vivo* studies [34]. In investigating the effects of nanoparticles shape on biodistribution, has been suggested that anisotropically shaped nanoparticles can avoid bioelimination better than spherical nanoparticles [29]. Also, biocompatibility is one of the main reasons that makes MNPs interesting for biomedical applications. For that reason, toxicity of MNPs is observed by cytotoxicity assays [35]. However, the cytotoxicity data of MNPs is difficult to compare since the toxic effects of MNPs are influenced by many parameters such as distribution, surface coating, and magnetic properties [36]. The most used techniques to estimate toxicity of magnetic nanoparticles *in-vitro* assays for cell viability, proliferation and differentiation are the MTT assay of mitochondrial function, the LDH assay of cell membrane integrity and immunochemistry markers for apoptosis/necrosis. A systematic and thorough quantitative *in-vivo* analysis of the pharmacokinetics, which examines adsorption, distribution, metabolism, and excretion of MNPs, can lead to improvements in design of biocompatible MNPs, a better understanding of nanoparticle non-specificity toward tissues and cell types, and assessments of basic distribution and clearance that serve as the basis to understand their activity and potential toxicity [37, 38]. Size and morphology of MNPs is usually performed by transmission electron microscope (TEM), where an average size of the nanoparticles was 9.1±2.2 nm with narrow size distribution (Figure 2) [39].

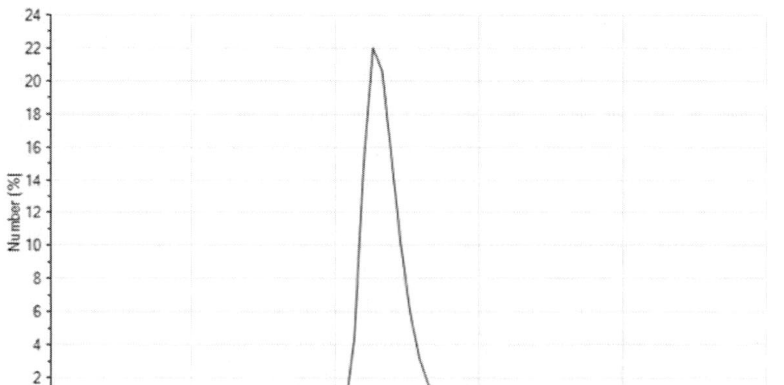

Figure 1. Size distribution of maghemite nanoparticles in aqueous medium.

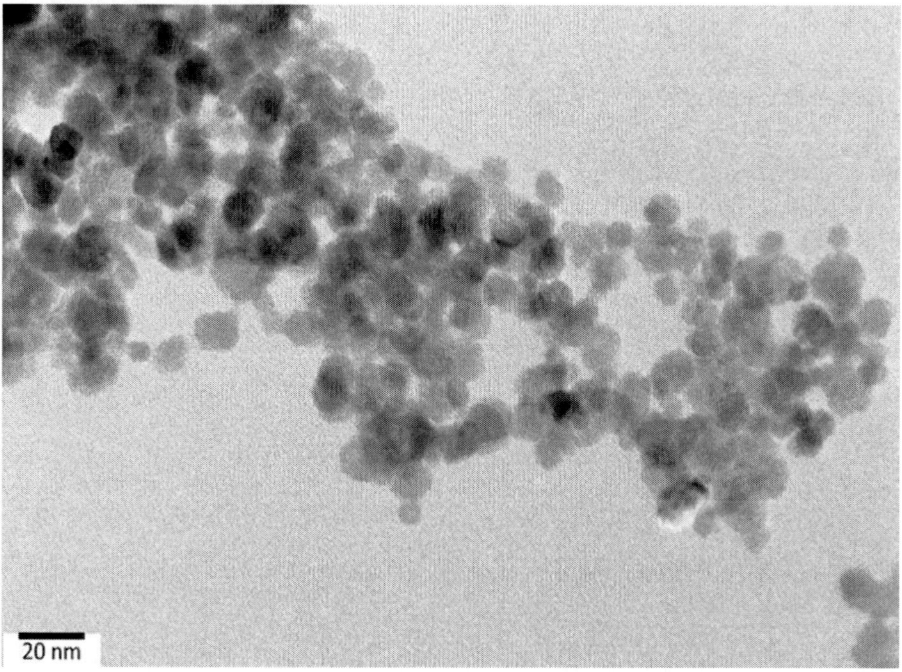

Figure 2. TEM image of maghemite nanoparticles coated with chitosan.

2.2. Magnetite, Maghemite, Iron-Based Metal Oxides

MNPs for the use in biomedical applications are desired to exhibit superparamagnetic properties [40] as shown in Figure 3. In this case superparamagnetic iron oxide nanoparticles (SPIONs) are obtained.

The vibrating sample magnetometer (VSM) is one of the most popular techniques in magnetometry. It is used to measure the magnetic moment m and the magnetic dipole moment $j = \mu_0 m$, where μ_0 is a magnetic constant, which is the value of magnetic permeability in a classical vacuum and appears only in SI units, in the presence of a static or slowly charging external magnetic field [41]. For all magnetizable materials, the VSM results are displayed as hysteresis curves.

The magnetization curve of bare maghemite nanoparticles revealed a typical saturation magnetization of a value 60 emu/g, approximately. It is expected that coating of the MNPs would lower their saturation magnetization, e.g., the magnetization of maghemite nanoparticles coated with chitosan was found to be equal to 40 emu/g.

Figure 4 showed that the chitosan surface functionalization of maghemite nanoparticles was achieved, but the difference in magnetization between uncoated and maghemite coated with chitosan appeared.

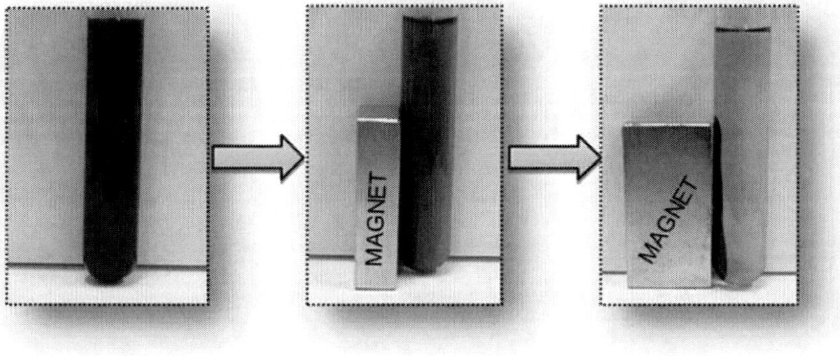

Figure 3. MNPs suspended in a liquid attracted to a magnet.

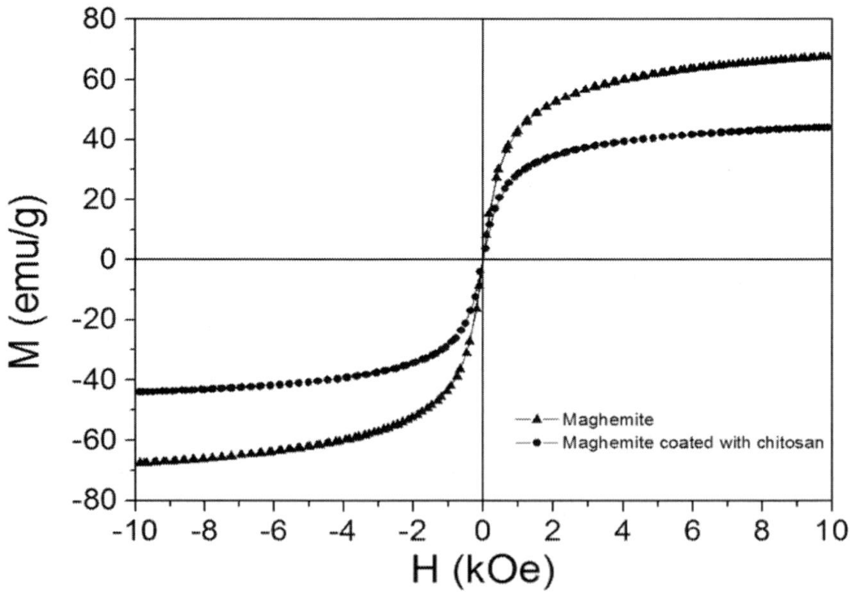

Figure 4. VSM magnetic measurements for maghemite and maghemite coated with chitosan.

SPIONs, most used as magnetic carriers, are typically small particles, composed of either magnetite (Fe_3O_4) or maghemite (γ-Fe_2O_3) core coated with a biocompatible organic/inorganic or polymeric coating. Both maghemite and magnetite are traditionally feromagnetic in nature. Feromagnetic materials have permanent magnetic moments. However, as their size decrease to 30 nm or lower, they lose their permanent magnetism and become superparamagnetic [42]. Therefore, iron oxide nanoparticles are suitable for biological application for the following reasons: the superparamagnetic nature implies that the particles will not be attracted to each other and can avoid aggregation [40]; Iron (Fe) is a naturally occurring metal in the human body, and hence iron-containing nanoparticles are biocompatible; have lower sedimentation rates (high stability), and improved tissular diffusion [43]. In addition, applications in biology and medical diagnosis and therapy require the magnetic particles to be stable in water at neutral pH and physiological salinity. Moreover, for *in-vivo* biomedical applications, MNPs must be made of nontoxic and nonimmunogenic material, with particle sizes small enough to remain in the circulation after injection and to pass through the capillary systems of organs and tissues. They also must have a high magnetization so that their movement in the blood can be controlled with a magnetic field [43]. However, biocompatibility is dependent on both cell type and MNPs surface chemistry, e.g., magnetic manganese oxide nanoparticles coated with a silica shell resulted in achieving better water stability at high concentrations and superior biocompatibility [26].

3. Synthesis of Magnetic Nanoparticles

In recent years, many different synthesis procedures of MNPs have been reported [34]. Some of them are one-step, while others are multi-step procedures, where advantages and disadvantages are presented [22]. However, major importance is to obtain shape and size controlled, highly stable and monodisperse MNPs, where the synthesis procedure should be relatively simple, inexpensive and reproducible [44, 45]. Summary comparison of various MNPs synthesis methods is shown in Table 1, where detailed description of each method is hereinafter.

Table 1. Different synthesis of MNPs along with their reaction conditions, advantages, and disadvantages

Synthetic method	Reaction temperature	Reaction period	Solvent	Size distribution	Advantages	Disadvantages	References
Coprecipitation	20-90 °C	Minutes	Water	10-200 nm	Very simple and rapid synthesis, easy control of particle size and composition	Wide particle size distribution, adjustment of pH value of the reaction mixture	[46]
Microemulsion	20-50°C	Hours	Organic compound	3.5-10 nm	Easy, highly reproducible, provides control of size, shape and colloidal stability	complicated, surfactants are difficult to remove, low productivity	[47, 48]
Hydrothermal synthesis	200-300 °C	Hours, days	Water-ethanol	12-40 nm	Good crystallinity and high saturation magnetization of NPs, low cost of synthesis, potential possibility of large scale production	Sharp reaction conditions, unwanted oxidation at the surface of metal NPs	[49, 50, 34]
Sol-gel method	70-400 °C	Hours	Organic compound	20-200 nm	Simple, low temperatures of synthesis, pure, monodisperse particles	Large size particles, high pressure, product usually contains sol-gel matrix components on their surfaces	[51, 52, 53, 50]
Thermal decomposition	120-320 °C	Hours,	Organic compound	4-20 nm	Narrow size distribution of the NPs, suitable for large scale production, good size control and good crystallinity	Complicated, inert atmosphere, process do not require a size-selection procedure and NPs are generally only dissolved in nonpolar solvents	[54, 46]

3.1. Coprecipitation

The most reported approach is by coprecipitation of Fe^{2+} and Fe^{3+} chloride, at pH 9-14, under non oxidising conditions. The balance of these requirements pH, type of salt, Fe^{2+} and Fe^{3+} ratio and ionic strength are reported to control the shape, size and composition of the SPIONs [55]. Furthermore, there are two main methods for the synthesis in solution of magnetic spherical nanoparticles. In the first, ferrous hydroxide suspension is partially oxidized with different oxidizing agents. The other method consists in ageing stoichiometric mixtures of ferrous and ferric hydroxides in aqueous media, yielding spherical magnetite nanoparticles. Furthermore, it has been shown that by adjusting the pH and the ionic strength of the precipitation medium, it is possible to control the mean size of the particles [56]. The size decreases as the pH and the ionic strength in the medium increases [57]. One important note in synthesis of MNPs is that magnetite (Fe_3O_4) nanoparticles are not very stable under ambient conditions and easily oxidized to maghemite (γ-Fe_2O_3) or dissolved in an acidic medium [17]. Unfortunately, particles prepared by the co-precipitation are likely to be polydisperse, due to two stages which are involved in the process and should be well-controlled to obtain monodisperse iron oxide MNPs. The crucial factors with high impact on size, shape and composition of MNPs prepared by coprecipitation are the type of used salts (e.g., chlorides, sulfates, nitrates), the Fe^{2+}/Fe^{3+} ratio, the reaction temperature, the pH value and ionic strength of the medium. When the ratio of Fe^{2+}/Fe^{3+} ions was increased it has been observed an increase in the size of Fe_2O_3 nanoparticles from 9 to 37 nm. Also, the magnetic properties of Fe_3O_4 nanoparticles can be controlled by adjusting the molar ratio of Fe^{2+}/Fe^{3+} ions. The suitable pH for the rapid formation of Fe_3O_4 to Fe_2O_3 is attained by the addition of excess amounts of the base and when using NH_4OH instead of NaOH for precipitation, a better crystallinity, higher saturation magnetization and smaller size can be observed [58]. Recently, SPIONs coated with dextran were prepared by chemical coprecipitation for *in-vivo* and *in-vitro* tissue distribution, where mean diameter was about 100 nm [59]. In terms of simplicity of the synthesis method, coprecipitation is the preferred route, where the reaction temperature and time is lower than in other methods such as thermal decomposition and hydrothermal synthesis. In addition, water as a solvent is environmental friendly. However, it is crucial that the particles obtained with the coprecipitation have large particle size distribution and sometimes, aggregation and poor crystallinity results in low saturation magnetization values [34].

3.2. Microemulsion

The water-in-oil (W/O) microemulsion is a thermodynamically stable isotropic dispersion of two immiscible liquids, where reverse micelles are tiny drops of water stabilized in a hydrophobic liquid phase surrounded by a monolayer of surfactant molecules in the continuous hydrocarbon phase [55, 45, 44]. The size of the reverse micelle is determined by the molar ratio of water to surfactant. Moreover, by the addition of solvent, such as acetone or ethanol, to the microemulsions, the precipitate can be extracted by filtering or centrifuging the mixture. In this context, a microemulsion can be used as a nano-sized "reactor" for the formation of nanoparticles. This process has been applied in generating magnetic iron oxide nanoparticles with size less than 10 nm [47]. Likewise, it is reported that maghemite nanoparticles with monolayer of oleylamine (or oleic acid) synthesized by microemulsion show a narrow size distribution of 3.5 ± 0.6 nm [48].

3.3. Hydrothermal Synthesis

This method is based on a general phase transfer and separation mechanism occurring at the interfaces of the liquid, solid and solution phases present during the synthesis in aqueous or organic systems. Hydrothermal synthesis, also called solvothermal, has been shown to produce iron oxide nanoparticles with narrow size distribution, where the synthesis is performed in aqueous media in reactors or autoclaves at pressure higher than 13.8 MPa and temperature above 200 °C. It is also relatively cost-effective method with high yield of products and excellent particle crystallinity with controllable size and good morphology of particles. The control of nanoparticle formation, size and distribution can be achieved by choosing the appropriate mixture of solvents and varying parameters such as temperature, pressure and reaction time [50]. Although hydrothermal technique is very versatile, one of the main drawbacks of this method is the slow reaction kinetics [58]. Under hydrothermal conditions the magnetite nanoparticles coated with chitosan were formed as a potential MRI contrast agents with narrow size distribution and average diameter of 17,12 nm and 57,40 emu/g of saturation magnetization [49].

3.4. Sol-Gel Method

Sol-gel technology has been extensively employed for the fabrication of a wide range of materials and has been found to be effective for dispersing small metal and oxide particles. Process is based on hydroxylation and condensation of molecular precursors in solution, originating a ″sol″ of nanoparticles. Further condensation and inorganic polymerization lead to a three-dimensional metal oxide network denominated wet ″gel″ [17]. Significantly, it is performed at low temperatures and can be used to synthesize pure, stoichiometric, monodisperse and large size nanoparticles [50]. Based on the conditions of preparation process, the size of the particles depend on three parameters, dilution of the solution, concentration of ligand (citric acid) and annealing temperature of the dried gel [51].

3.5. Thermal Decoposition

Up to date, two different approaches have been used to produce MNPs by thermal decomposition. Firstly, thermal decomposition of metal carbonyl precursors followed by an oxidation step using air, or oxidation by using an oxidant at elevated temperatures. The second is decomposition of precursors with cationic metal center in the absence of reducing agents [17]. On the other hand, decomposition of iron precursors in the presence of hot organic surfactants has given samples with good size control, narrow size distribution and good crystallinity of individual and dispersible magnetic iron oxide nanoparticles [56]. Very recently, 9 nm and 16 nm sized porous iron oxide nanoparticles with encapsulated doxorubicin were synthesized via thermal decomposition to be used as MRI contrast agent and as a drug delivery vehicle [60] and for localized, triggered drug delivery [61].

4. Surface Coating, Stabilization and Functionalization

One of the iron oxide nanoparticle characteristics are hydrophobic surfaces with a large surface area to volume ratio, which is causing the tendency to agglomerate as a result of their high surface energy [55, 62]. Surface modification strategies needed to change nanoparticle surface from hydrophobic to hydrophilic [63]. The coating renders the particles biocompatible, prevents them from oxidation in suspension, isolates them from each other, avoids the formation of magnetic aggregates, thus improving their dispersibility and colloidal stability, reducing toxicity, and serves as a support for the chemical bonding or physical adsorption of biomolecules on the surface of the particle [18, 34]. However, magnetic iron oxide nanoparticles are usually coated with organic molecules including surfactants, polymers and biomolecules or inorganic layer such as silica and metal oxide. Generally, surfactants can be chemically anchored or physically adsorbed on MNP core to form a single or double layer as illustrated in Figure 5. Polymers containing functional groups, such as amino, carboxylic acids, thiol, phosphates, and sulfates, can bind to the surface of MNPs [29, 44]. Furthermore, covalent linkages are strong and stable bonds, which can be specifically formed between functional groups, found on the nanoparticle surface and conjugated ligands.

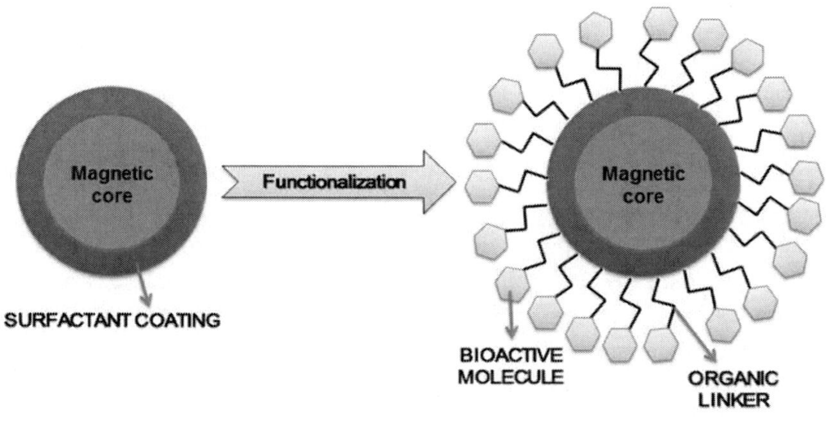

Figure 5. Illustration of the binding molecule on MNP.

The most commonly used coatings to stabilize the MNPs and for other biological applications are oleic acid [3,4,25,36,64], citric acid [65], polyethylene glycol (PEG) [2,5,6,13,61,66], dextran [59], chitosan [39,67,14], starch [68], polyvinyl alcohol (PVA) [69], human serum albumin (HSA) [10,60], silica [7,12,15,24], silane [70], gold [16].

Coating process involves multiple steps, and usually results in coating thickness, which can drastically affect the magnetic properties as significant loss in the magnetization of the magnetic core [29]. This decrease in magnetization negatively influences the magnetic targeting ability of the carrier system *in-vivo* [71].

Practically, it is worthy that in many cases the coatings not only stabilize the magnetic iron oxide nanoparticles, but can also be used for further functionalization [46]. Accordingly, surface functionalization is a key process towards reconstructed derivatives suitable for certain demands.

5. Cytotoxicity Studies

Beside particle size, surface charge and protein binding capacity, toxicity of the nano-particles is the most important property for their use in a specific nano-carrier system *in-vivo* application. It is important to evaluate interactions of MNPs system with the body during the therapy. The tolerance of many developed MNPs is expectable. Toxicity of the MNPs depends on numerous factors including the dose, chemical composition, method of administration, size, biodegradability, solubility, pharmacokinetics, biodistribution, surface chemistry, shape, and structure [72]. Because of the great variety in types of MNPs, cells and incubation protocols, it is impossible to make any conclusions regarding the safety of MNPs for bio-medical implication. At nontoxic concentrations of MNPs, several of them might not be well suited for their original purpose, so it is important for the cytotoxicity assessments to evaluate MNPs efficiency, as well as the deep understanding of the toxicity pathways of various MNPs [64]. In the case when MNPs shall be used for cancer therapeutic applications, a high cytotoxicity on cancer cells and low cytotoxicity on normal cells is required; e.g., chitosan-ferroferric oxide doped cadmium telluride quantum dot nanoparticles possess higher growth inhibition property for cancer cell than for normal cell [67].

It is also very important to investigate the toxicity of the non-drug coated MNPs. The reason is accumulation of the MNPs in the body after drug release. This is especially important when slowly or non-degradable particles are used to prevent eventually chronic inflammatory reactions [73].

Among iron oxide based MNPs only maghemite (γ-Fe_2O_3) and magnetite (Fe_3O_4) are suitable for biomedical applications, since they fulfill the requirements, such as their superior biocompatibility with respect to other magnetic materials. Toxicity of the MNPs can be associated with the iron oxide core of MNPs. The iron dose of the MNPs is many times lower than body iron levels. Injected iron would therefore be easily metabolized and regulated by the body's normal physiological iron homeostatic mechanisms.

Biodistribution, elimination, and oxidative stress studies have shown that MNPs have a very low toxicity, and minimal effect on the body normal homeostasis. The injected iron does not cause long-term changes in liver enzyme levels or oxidative stress, and most of the changes are only transient [71]. Even with the incorporation of higher levels of iron into the endosome, the iron is not released into the cytoplasm as a free radical form, which would be toxic to the cells and increased cell death [27]. Anyway, coating MNPs with a biocompatible polymer is necessary when high concentrations (> 100 µg/mL) of MNPs are used [70] while at such high concentrations toxicity of the MNPs increases significantly above a certain administration level. The concentrations, needed for drug delivery are usually below the toxic level for suitable coated MNPs [74]. For *in-vivo* applications, the magnetic particles must be coated with a biocompatible polymer during or after the synthesis process to prevent the formation of large aggregates, changes from the original structure and biodegradation when exposed to the biological system, as well [43]. MNPs coated with longer-chain triblock copolymers (5 and 15 kDa) were found to be nontoxic to endothelial (HUVEC), epithelial (HRPE), and tumor (PC3 and C4-2) cells. The inhibitory dose that resulted in 50% cell viability, the ID50, was found to be above, and often far above, 5 mg of MNPs per milliliter of cell media, which is at least 10× higher than any MNP concentration foreseen to be used in biomedical applications [75]. At HSA coated iron oxide nanoparticle (HINP) no inhibitive effects of HINPs against cell growth were found. As a matter of fact, even slightly increased viability was observed at higher Fe concentrations [10]. Usually, when high cytotoxicity is deteced at high MNPs concentrations, this may be attributed to the toxicity of the drug bounded to the MNPs [20]. Studies of the toxicity of different formulated MNPs on the cells have been performed.

Citotoxicty profile of the PEG-modified SPIONs, with specific shape and size, prepared inside the aqueous cores of AOT/*n*-hexane reverse micelles, on human dermal fibroblasts as measured by standard 3-(4,5-dimethylthiazol-2-yl)-2,5-diphenyltetrazolium bromide assay showed that the particles are nontoxic and may be useful for relevant drug delivery and other biomedical applications [5]. Low toxicity was determined for coated and uncoated iron oxide nanoparticles, prepared from different precursors (iron acetate, iron pentacarbonyl, decalin, PEG (poly(ethylene glycol)), EG (ethylene glycol), PVA (poly(vinyl alcohol)), β-cyclodextrin, and distilled water) using sonochemical method. Toxicity studies were carried out to determine the effect of the as-prepared coated and uncoated iron oxide nanoparticles on human cell viability at increasing concentrations and exposure time [13].

Biocompatible coatings on MNPs express lower toxicity also due to the fact that lower adsorption sites for serum proteins which could bind to the surface of the MNPs are available [76].

The importance of the surface area optimization with proper coating showed the study where citric acid and alginic acid surfactants were found to be significantly less toxic than starch, decanoic acid and PEG [77]. Very low cytotoxicity was also detected for chitosan and poly L-lactic acid coated MNPs [78, 79].

In-vitro citotoxicity (MTT Assay) was studied for water-dispersible SPIONs, where iron oxide core nanoparticles were obtained by precipitation of iron salts in the presence of ammonia and provided β-cyclodextrin and pluronic polymer (F127) coatings. The so prepared MNPs are suitable for hyperthermia, magnetic resonance imaging (MRI) and drug delivery applications. Almost similar growth inhibitory effect for the with anti-cancer drug encapsulated MNPs was detected as for the pure anti-cancer drug in the cancer cells [23].

Good cytocompatibility based on the MTT cytotoxicity assay was determined for magnetite (Fe_3O_4) - loaded polymer micelles (denoted as "magnetomicelles"), produced by selfassembly of fluorine-containing amphiphilic poly(HFMA-g-PEGMA) copolymers with oleic acid modified Fe_3O_4 nanoparticles in an aqueous medium, which suggests their good compatibility and suitability as nanocarriers or probes in biomedical applications [64].

On the basis of the cytotoxicity tests rhodium (II) citrate-loaded maghemite nanoparticles or magnetoliposomes should be considered as suitable drug delivery systems, on both normal and carcinoma breast cell cultures, since they can act preferentially in tumor cells [80]. Similar, SPIONs and oxaliplatin (OHP) in-situ encapsulated in pectin cross-linked with Ca2þ forming 100–200 nm sized magnetically functionalized pectin nanocarriers, exhibited 10 folds higher cytoxicity on MIA-PaCa-2 (pancreas) cancer cell line than the equivalent concentration of free drug [81]. Similar, maghemite nanoparticles coated with citrate and rhodium (II) citrate showed enhanced cytotoxicity on breast carcinoma cells by up to 4.6 times when compared to free rhodium (II) citrate, which strengthens the case for its use in future breast cancer therapeutic applications [82].

A composite MNP formulation composed of an iron oxide nanoparticle core coated with β-cyclodextrin and pluronic polymer (F68) with curcumin loading (MNP-CUR) capability exhibited improved uptake in cancer cells, increased loss of mitochondrial membrane potential, and produced higher reactive oxygen species, suggesting it is extremely valuable in anticancer therapeutics [28]. In *in-vitro* evaluations the paclitaxel (TXL)-loaded SiMNP formulations demonstrated concentration-dependent cytotoxicity against tumor cells [24, 67].

6. Biomedical Applications of MNPs

There is a possibility of two types of biomedical applications of magnetic nanocarriers, coated with bioactive substances; *in-vitro* and *in-vivo*, where *in-vivo* applications may be further divided in diagnostic and therapeutic applications. MNPs properties, such as dimensions, stability in the environment, toxicity, surface chemistry and magnetic properties determine the type of application. As an example, SPION(Au) core(shell) nanoparticles with narrow size distribution and different shell thicknesses were synthesized with a method of synthesis which not only allows authors to tune the optical response by varying the shell thickness but also to tune the magnetization of these particles to suit different biomedical applications [16].

6.1. In-Vitro Applications

There is a wide application of MNPs in medical diagnostics, such as magnetic resonance imaging (MRI), where MNPs play a role as contrast agents and may be used for *in-vivo* and *in-vitro* cellular and molecular imaging [83].

For *in-vitro* application MNPs of bigger size may be used than for *in-vivo* applications. However, using MNPs instead magnetic microparticles in magnetic solid-phase extractions suspensions that are stable against sedimentation in the absence of an applied magnetic field may be prepared.

In bioseparation applications MNPs have many superior characteristics for separation and purification of cells and biomolecules in bioprocesses compared to those of the conventional micrometer-sized resins or beads such as good dispersability, the fast and effective binding of biomolecules, and reversible and controllable flocculation [83].

Another possible application of MNPs is their use in biocatalysis, where the possibility of easy separation of heterogeneous catalyst from the reaction mixture plays a significant role. This is very important especially for small size expensive biocatalysts, which can be separated simple with magnetically driven separation methods. Another variation in MNPs use in catalysis was their role as carriers for metal catalysis in different types of reactions, such as carbon-carbon cross-coupling reactions, hydroformylation [84], hydrogenation, and polymerization [85] reactions.

6.2. In-Vivo Applications

A new class of pharmaceuticals, the so called magneto-pharmaceuticals, are needed for diagnostic applications. MNPs as commercial products are used as contrast agents in MRI for location and diagnosis of brain and cardiac infarcts, liver lesions or tumors [56].

In the case when SPIONs are used less amount is required to be loaded in human body compared with the use of paramagnetic MNPs. This is very important because an important amount of particles was found to be concentrated in the main filters of the body, the liver and spleen, whereas a much lower quantity of particles was found in the lungs of some animals. Despite the inert surface of the material, the size of the particles was large enough to be recognized by the macrophages of these organs, which captured and retained them, eventually excreting them after several days [18]. The level of iron in the spleen after 3 weeks corresponded to 25 % of injected dose.

A relatively short blood half-life was detected for most iron oxides so their primary application is for imaging of the liver, spleen, and gastrointestinal tract. On the contrary, biodistribution studies performed in mice for surface-modified iron oxide nanoparticles showed long blood circulation times, and may be useful for imaging of the vascular compartment (magnetic resonance angiography), imaging of lymph nodes, perfusion imaging, receptor imaging, and target-specific imaging [5].

Use of MNPs in hyperthermia and drug delivery belongs to their therapeutic applications.

In hyperthermia, where tumor cells are destroyed with temperatures between 41–42 ˚C, heating should be restricted to the tumor area. The use of MNPs of uniform shape is preferred instead magnetic microparticles because and size and shape of the particles is essential for high power absorbance at tolerable AC magnetic fields [86-88].

A relative new technology is the use of MNPs for delivery of drugs or antibodies to the organs or tissues altered by diseases. There has been a lot of research done lately in this field [56, 89, 90]. MNPs with functionalized surface, where bioactive molecules are attached, in combination with an external magnetic field are able to transport drugs to a specific site in the human body. With this technique side effects can be eliminated and the required dosage can be reduced. Additionally, two goals can be reached; the targeting of the drug to the desired area in the body to reduce side effects to other organs and controlled release of the drug to avoid the classical overdosing/underdosing cycle [35]. Two types of MNPs coating exist; covalent coupling and physical absorption of bioactive substances and both types of obtained products showed high potentials for drug delivery [37]. For *in-vivo* biomedical applications MNPs must be made of a nontoxic and nonimmunogenic material, with particle sizes small enough to remain in the circulation after injection and to pass through the capillary systems of organs and tissues avoiding vessel embolism [43].

Drug delivery on the base with MNPs has been investigated for different types of therapeutic treatments. AMT-functionalized MNPs were characterized *in-vivo* to a kainic acid model of epilepsy. The affinity of the AMT-MNPs for epileptic tissues was confirmed [91]. A successfull antitumor activity was retained for Dox loaded magnetite nanocrystals on the surface (Fe_3O_4-MSN) when injection doses of Dox were 2 and 4 mg/kg mice [12]. It was confirmed that multifunctional quaternary nanoparticles designed as a targeting probe for cancer cells could potentially provide an approach for diagnosis, drug delivery, and real-time detection on drug treatments of the intracellular processes [67].

7. Drug Loading and Release

Traditional drug delivery systems are limited by their passive kinetics of drug release which means that drugs inside traditional delivery vehicles are continuously and uncontrollably released [92, 93]. Instead, using MNPs, quantitative amounts of drug release can be controlled precisely by alternating current magnetic field (AMF), which serves as a trigger, to promote drug release in spatio-temporally controlled manner where the drug is released in an on-demand manner [63].

In principle, the magnetically sensitive drug carrier is delivered through a pathway either systematically or semi-systematically providing carrier particles throughout the vasculature in the case of arterial or venous delivery or throughout the organ's vasculature in the case of delivery near the organ. Once delivered, a magnetic field is used to interact with particles [42]. Depending on the formulation, MNPs may also be applied with penetration through the skin. Once the MNPs are located onto the cellular surface, fast internalization begins, followed by degradation of the linkage between a MNP and therapeutic within body fluids. Because of their higher endocytotic potential malligant cells are more prone to internalization of MNPs than normal cells [94].

Surface properties of MNPs which depend on the substance used for their stabilization are essential for their behavior *in-vivo*. Attachment of the targeting ligand, chlorotoxin, to SPIONs coated with a copolymer of polyetheyleneimine, PEG, and chitosan (NP–CP–PEI) increased transfection efficiency 3-fold [95]. It was found that more than 50 % of the PEG-coated (P-SPIONs) were still circulating in mice blood after 60 min to finally totally disappear after 360 min, which enables us to be optimistic about the use of those maghemite nanoparticles for magnetic drug targeting [66]. The t1/2 for P-SPIONs was estimated to be 76 ± 6 min. P-SPIONs are too large to be rapidly removed through renal clearance because the cutoff size for renal excretion is approximatively 5.5 nm according to recent research using quantum dots [96]. Moreover, they are too small to be quickly sequestered by phagocytic cells of the liver and spleen [97]. Enhanced, selective magnetic brain tumor targeting (t ¼ 1 h) of PEG-MNPs (12 mg Fe/kg) was confirmed in 9L-glioma tumors, with up to 1.0 % injected dose/g tissue nanoparticle delivery achieved - a 15-fold improvement over targeted D (0.07 % injected dose/g tissue). Drug-loaded PEG-MNPs should be further developed and concurrent optimization of the magnetic targeting strategy shall be utilized [68].

It is known that when NPs with high colloidal stability under physiological conditions and size smaller than around 200 nm are injected intravenously, parts of maghemite nanoparticles accumulate preferentially at tumor sites through their enhanced permeability and retention (EPR) effect in passive targeting. Magnetic targeting of MNP-CUR facilitated an enhanced delivery of curcumin to cancer cells [28].

To determine their drug delivery application, doxorubicin (base (Dox)) as a model anticancer cancer drug was loaded and studied the anti-proliferative effect of the drug-loaded MNPs in the MCF-7 breast cancer cell line. The oleic acid (OA) coat around the iron-oxide core acted as a reservoir for loading of hydrophobic drug. The sustained release of Dox from the MNPs over several days likely occured through two mechanisms; drug release from the surface of the MNPs and through diffusion from OA layers due to concentration gradient with respect to the surrounding environment [2].

In-vitro studies were performed to elucidate what is the role of nanoparticles in Dox internaction and intracellular distribution. Dox and Dox loaded human serum albumin (HSA) coated iron oxide nanoparticle (HINP) D-HINPs were incubated with 4T1 cells, and monitored the movement of Dox using fluorescence microscopy. Dox was initially distributed in the cytoplasm (at 5 min), but later concentrated in the nuclei, where it is expected to exert therapeutic power. For D-HINPs, however, such a process was accelerated, as intensive Dox signals were found in the nuclei merely 5 min after the start of incubation. On the other hand, the HINPs were found exclusively in the cytoplasm, which is not surprising considering their size. These observations suggest that HINPs, at least at early time points, played a significant role in transporting Dox [10].

pH controlled drug release was demonstrated with successfully coated SPIONs with chitosan and gallic acid (FCG). This controlled-release study showed that the release rate of gallic acid from FCG depends on the pH of the media and type of the host; when pH was 7.4, the release rate was slower than that at pH 4.8. It is also apparent that gallic acid was released in a controlled manner [14].

pH-sensitive drug release kinetics was also demonstrated at SPION-Dox release analysis which showed fast Dox release at pH 5.5 and 6.4 (as in endosomal environment) and slow release at pH 7.4 (physiological condition) [98].

CONCLUSION

Due to their special properties, such as ultra-fine size, functionality, stability, biocompatibility and superparamagnetic properties MNPs have been lately considered for a number of *in-vitro* and *in-vivo* biomedical applications. Their size should be much smaller than cell and depends on the synthetic route used for their preparation. MNPs should be coated with different types of substances, such as organic molecules, including surfactants, polymers and biomolecules or inorganic layer such as silica and metal oxide to make them biocompatible, prevent them from oxidation in suspension, isolate them from each other, avoid the formation of magnetic aggregates, thus improve their dispersibility and colloidal stability, reduce toxicity, and serving as a support for the chemical bonding or physical adsorption of biomolecules on the surface of the particle. Very low or no cytotoxicity and minimal effect on the body homeostasis has been found for MNPs thus they may be used as an excellent tool for targeted drug delivery with controlled drug release.

ACKNOWLEDGMENTS

The authors would like to thank dr. Slavko Kralj from IJS, Ljubljana, Slovenia for his help in operating TEM. This work was entirely supported by the Slovenian Research Agency (PhD researcher fellowship contract No. 1938/FKKT-2010) and financially supported by Slovenian Research Agency through Project J2-4232.

REFERENCES

[1] Chomoucka J.; Drbohlavova J.; Huska D.; Adam V.; Kizek R.; Hubalek J. Magnetic nanoparticles and targeted drug delivering. *Pharmacological research* 2010, 62, 144-149.

[2] Yallapu M.M.; Foy S.P.; Jain, T.K.; Labhasetwar V. PEG-functionalized magnetic nanoparticles for drug delivery and magnetic resonance imaging applications. *Pharmaceutical research* 2010, 27, 2283-2295.

[3] Shin J.; Anisur R.M.; Ko M.K.; Im G.H.; Lee J.H.; Lee I.S. Hollow manganese oxide nanoparticles as multifunctional agents for magnetic resonance imaging and drug delivery. *Angewandte chemie* 2009, 48, 321-324.

[4] Jain T.K.; Morales M.A.; Sahoo S.K.; Leslie-Pelecky D.L.; Labhasetwar V. Iron oxide nanoparticles for sustained delivery of anticancer agents. *Molecular pharmaceutics* 2005, 2 (3), 194-205.

[5] Gupta K.A.; Wells S. Surface-modified superparamagnetic nanoparticles for drug delivery: preparation, characterization, and cytotoxicity studies. *Ieee transactions on nanobioscience* 2004a, 3, 66-73.

[6] Gupta K.A.; Curtis A.S.G. Surface modified superparamagnetic nanoparticles for drug delivery: interaction studies with human fibroblasts in culture. *Journal of materials science: Materials in medicine* 2004b, 15, 493-496.

[7] Kim J.; Kim H.S.; Lee N.; Kim T.; Kim H.; Yu T.; Song I.C.; Moon W.K.; Hyeon T. Multifunctional uniform nanoparticles composed of a magnetite nanocrystal core and a mesoporous silica shell for a magnetic resonance and fluorescence imaging and for drug delivery. *Angewandte chemie* 2008, 47, 8438-8441.

[8] Chertok B.; Moffat B.A.; David A.E.; Yu F.; Bergemann C.; Ross B.D.; Yang V.C. Iron oxide nanoparticles as a drug delivery vehicle for MRI

monitored magnetic targeting of brain tumors. *Biomaterials* 2008, 29, 487-496.

[9] Silva-Freitas E.L.; Carvalho J.F.; Pontes T.R.F.; Araujo-Neto R.P.; Carrico A.S.; Egito E.S.T. Magnetite content evaluation on magnetic drug delivery systems by spectrophotometry: A technical note. *American association of pharmaceutical scientists* 2011, 12 (2), 521-524.

[10] Quan Q.; Xie J.; Gao H.; Yang M.; Zhang F.; Liu G.; Lin X.; Wang A.; Eden H.S.; Lee S.; Zhang G.; Chen X. HSA coated iron oxide nanoparticles as drug delivery vehicles for cancer therapy. *Molecular pharmaceutics* 2011, 8 (5), 1669-1676.

[11] Guo S.; Li G.; Zhang L.; Li Z.; Wang E. Monodisperse mesoporous superparamagnetic single-crystal magnetite nanoparticles for drug delivery. *Biomaterials* 2009, 30, 1881-1889.

[12] Lee J.E.; Lee N.; Kim H.; Kim J.; Choi S.H.; Kim J.H.; Kim T.; Song I.C.; Park S.P.; Moon W.K.; Hyeon T. Uniform mesoporous dye-doped silica nanoparticles decorated with multiple magnetite nanocrystals for simultaneous enhanced magnetic resonance imaging, fluorescence imaging, and drug delivery. *Journal of American chemical society* 2009, 132, 552-557.

[13] Bolden N.W.; Rangari V.K.; Jeelani S.; Boyoglu S.; Singh S.R. Synthesis and evaluation of magnetic nanoparticles for biomedical applications. *Journal of nanoparticles* 2013, 1-9.

[14] Dorniani D.; Hussein M.Z.B.; Kura A.U.; Fakurazi S.; Shaari A.H.; Ahmad Z. Preparation of Fe_3O_4 magnetic nanoparticles coated with gallic acid for drug delivery. *International journal of nanomedicine* 2012, 7, 5745-5756.

[15] Hiroki T.; Taira S.; Katayanagi H.; Moro Y.; Shigeoka D.; Kimura S.; Mashino T.; Ichiyanagi Y. Functional magnetic nanoparticles for use in a drug delivery system. *Journal of physics* 2010, 1-4.

[16] Gangopadhyay P.; Gallet S.; Franz E.; Persoons A.; Verbiest T. Novel superparamagnetic core(shell) nanoparticles for magnetic targeted drug delivery and hyperthermia treatment. *Ieee transactions on magnetics* 2005, 41 (10), 4194-4196.

[17] Tran P.H.; Tran T.T.; Vo T.T.; Lee B. Promising iron oxide-based magnetic nanoparticles in biomedical engineering. *Archives of pharmacal research* 2012, 32 (3), 444-450.

[18] Fernández-Pacheco R.; Valdivia G. Magnetic nanoparticles for local drug delivery using magnetic implants. *Micro and nano technologies in*

bioanalysis, Lee J.W., Foote R.S.; Humana press: Oak Ridge, TN, USA, 2009; 544, 559-569.

[19] Neuberger T.; Schöpf B.; Hofmann H.; Hofmann M.; Rechenberg B. Superparamagnetic nanoparticles for biomedical applications: possibilities and limitations of a new drug delivery system. *Journal of magnetism and magnetic materials* 2005, 293, 483-496.

[20] Wichita state university. Composite magnetic nanoparticle drug delivery system, Asmatulu R.; Misak H.; Yang S.; Wooley P. US2012/0265001A1. US patent. 2012-10-18.

[21] Alexiou C.; Schmid R.J.; Jurgons R.; Kremer M.; Wanner G.; Bergemann C.; Huenges E.; Nawroth T.; Arnold W.; Parak F.G. Targeting cancer cells: magnetic nanoparticles as drug carriers. *European biophysics journal* 2006, 35, 446-450.

[22] McBain S.; Yiu H.H.; Dobson J. Magnetic nanoparticles for gene and drug delivery. *International journal of nanomedicine* 2008, 3(2), 169-180.

[23] Yallapu M.M.; Othman S.F.; Curtis E.T.; Gupta B.K.; Jaggi M.; Chauhan S.C. Multi-functional magnetic nanoparticles for magnetic resonance imaging and cancer therapy. *Biomaterials* 2011, 32 (7), 1890-1905.

[24] Auzenne E.A.; Seeney C.E.; Hamir A.N.; Klostergaard J. Anti-tumor activity of drug-loaded magnetic nanoparticles. *Ieee transactions on magnetics* 2013, 49 (1), 336-342.

[25] Jain T.K.; Richey J.; Strand M.; Leslie-Pelecky D.L.; Flask C.A.; Labhasetwar V. Magnetic nanoparticles with dual functional properties: Drug delivery and magnetic resonance imaging. *Biomaterials* 2008, 29, 4012-4021.

[26] Kumar C.S.S.R.; Mohammad F. Magnetic nanomaterials for hyperthermia-based therapy and controlled drug delivery. *Advanced drug delivery reviews* 2011, 63, 789-808.

[27] Frank J.A.; Miller B.R.; Arbab A.S; Zywicke H.A.; Jordan E.K.; Lewis B.K.; Bryant L.H.; Blute J.W.M. Clinically applicable labeling of mammalian and stem cells by combining superparamagnetic iron oxides and transfection agents. *Radiology* 2003, 228 (2), 480-487.

[28] Yallapu M.M.; Othman S.F.; Curtis E.T.; Bauer N.A.; Chauhan N.; Kumar D.; Jaggi M.; Chauhan S.C. Curcumin-loaded magnetic nanoparticles for breast cancer therapeutics and imaging applications. *International journal of nanomedicine* 2012, 7, 1761-1779.

[29] Veiseh O.; Gunn J.W.; Zhang M. Design and fabrication of magnetic nanoparticles for targeted drug delivery and imaging. *Advanced drug delivery reviews* 2010, 62, 284-304.
[30] Chen D.H.; Liao M.H. Preparation and characterization of YADH-bound magnetic nanoparticles. *Journal of molecular catalysis* 2002, 16, 283-291.
[31] Sawant R.R.; Torchilin V.P. Multifunctional nanocarriers and intracellular drug delivery. *Current opinion in solid state and materials science* 2012, in press.
[32] Dutta R.K.; Sahu S. Development of oxaliplatin encapsulated in magnetic nanocarriers of pectin as a potential targeted drug delivery for cancer therapy. *Results in pharma sciences* 2012, 2, 38-45.
[33] Frey N.A.; Peng S.; Cheng K.; Sun S. Magnetic nanoparticles: synthesis, functionalization, and applications in bioimaging and magnetic energy storage. *Chemical society reviews* 2009, 38 (9), 2532-2542.
[34] Wahajuddin S.A.; Superparamagnetic iron oxide nanoparticles: magnetic nanoplatforms as drug carriers. *International journal of nanomedicine* 2012, 7, 3445-3471.
[35] Bucak S.; Yavuztürk B.; Sezer A.D. Magnetic nanoparticles: synthesis, surface modifications and application in drug delivery. Recent advances in novel drug carrier system. Sezer A.D.; InTech, Croatia, 2012; 165-200.
[36] Kai W.; Xiaojun X.; Ximing P.; Zhenqing H.; Qiqing Z. Cytotoxic effects and the mechanism of three types of magnetic nanoparticles on human hepatoma BEL-7402 cells. *Nanoscale research letters* 2011, 6:480.
[37] Liu G.; Gao J.; Ai H.; Chen X. Applications and potential toxicity of magnetic iron oxide nanoparticles. *Small* 2013, 9, 1533-1545.
[38] Wamocha H.L.; Misak H.E.; Song Z.; Chu H.Y.; Chen Y.Y.; Asmatulu R.; Yang A.Y.; Ho J.C. Cytotoxicity of release products from magnetic nanocomposites in targeted drug delivery. *Journal of biomaterials applications* 2011, 27, 661-667.
[39] Hojnik Podrepšek G.; Knez Ž.; Leitgeb M. Different preparation methods and characterization of magnetic maghemite coated with chitosan. *Journal of nanoparticle research* 2013, 15, 1-12.
[40] Markides H.; Rotherdam M.; El Haj A.J. Biocompatibility and toxicity of magnetic nanoparticles in regenerative medicine. *Journal of nanomaterials* 2012, 2012, 1-11.

[41] Svoboda J. Principles of material treatment by magnetic means. *Magnetic techniques for the treatment of materials*, Svoboda J.; Kluwer academic publishers, Dordrecht, 2004; 1-66.

[42] Abbott cardiovascular system inc. Magnetically sensitive drug carriers for treatment or targeted delivery, Davalian D.; Hossainy S.F.A.; Bright R.; Wan J.; Ludwig F.N. US2011/0196474A1. US patent. 2010-2-11.

[43] Nikiforov V.N.; Filinova E.Y. Biomedical applications of magnetic nanoparticles. *Magnetic nanoparticles,* Gubin S.P.; WILEY-VCH, Weinheim, 2009; 393-455.

[44] Lu A.H.; Salabas E.L.; Schüth F. Magnetic nanoparticles: synthesis, protection, functionalization, and application. *Magnetic nanoparticles* 2007, 46, 1222-1244.

[45] Gubin S.P.; Koksharov Y.A.; Khomutov G.B.; Yurkov G.Y. Magnetic nanoparticles: preparation, structure and properties. *Russian chemical reviews* 2005, 74 (6), 489-520.

[46] Wu W.; He Q.; Jiang C. Magnetic iron oxide nanoparticles: Synthesis and surface functionalization strategies. *Nanoscale research letters* 2008, 3, 397-415.

[47] Chin A.B.; Yaacob I.I. Synthesis and characterization of magnetic iron oxide nanoparticles via w/o microemulsion and Massart's procedure. *Journal of materials processing technology* 2007, 191, 235-237.

[48] Vidal-Vidal J.; Rivas J.; Lopez-Quintela M.A. Synthesis of monodisperse maghemite nanoparticles by the microemulsion method. *Colloids and surfaces* 2006, 288, 44-51.

[49] Haw C.Y.; Mohamed F.; Chia C.H.; Radiman S., Zakaria S.; Huang N.M.; Lim H.N. Hydrothermal synthesis of magnetite nanoparticles as MRI contrast agents. *Ceramics international* 2010, 36, 1417-1422.

[50] Sharifi I.; Shokrollahi H.; Amiri S. Ferrite-based magnetic nanofluids used in hyperthermia applications. *Journal of magnetism and magnetic materials* 2012, 324, 903-915.

[51] Akbar S.; Hasanain S.K.; Azmat N.; Nadeem M. Synthesis of Fe2O3 nanoparticles by new sol-gel method and their structural and magnetic characterizations. *Materials science* 2004, 1-19.

[52] Deng Y.H.; Wang C.C.; Hu J.H.; Yang W.L.; Fu S.K. Investigation of formation of silica-coated magnetite nanoparticles via sol-gel approach. *Colloids and surfaces A: Physicochemical and engineering aspects* 2005, 262, 87-93.

[53] Raja K.; Verma S.; Karmakar S.; Kar S.; Jerome Das S.; Bartwal K.S. Synthesis and characterization of magnetite nanocrystals. *Crystal research and technology* 2011, 5, 497-500.

[54] Cao X.; Zhang B.; Zhao F.; Feng L. Synthesis and properties of MPEG-coated superparamagnetic magnetite nanoparticles. *Journal of nanomaterials* 2012, 2012, 1-6.

[55] Gupta K.A.; Gupta M. Synthesis and surface engineering of iron oxide nanoparticles for biomedical applications. *Biomaterials* 2005, 26, 3995-4021.

[56] Tartaj P.; Morales M.P.; Veintemillas-Verdaguer S.; Gonzales-Carreno T.; Serna C.J. The preparation of magnetic nanoparticles for applications in biomedicine. *Journal of physics D: Applied physics* 2003, 36, 182-197.

[57] Indira T.K.; Lakshmi P.K. Magnetic nanoparticles – A review. *International journal of pharmaceutical sciences and nanotechnology* 2010, 3, 1035-1042.

[58] Faraji M.; Yamini Y.; Rezaee M. Magnetic nanoparticles: synthesis, functionalization, characterization, and applications. *Journal of Iranian chemical society* 2010, 7, 1-37.

[59] Wang L.; Zhang Y.; Li S.; Wang Y.; Wang K. Comparison of two kinds of magnetic nanoparticles *in vivo* and *in vitro*. *Journal of Huazhong University of science and technology* 2012, 32, 444-450.

[60] Xing R.; Bhirde A.A.; Wang S.; Sun X.; Liu G.; Hou Y.; Chen X. Hollow iron oxide nanoparticles as multidrug resistant drug delivery and imaging vehicles. *Nano research* 2013, 6, 1-9.

[61] Glover A.L.; Bennett J.B.; Pritchett J.S.; Nikles S.M.; Nikles D.E.; Nikles J.A.; Brazel C.S. Magnetic heating of iron oxide nanoparticles and magnetic micelles for cancer therapy. *IEEE Transactions on magnetics* 2013, 49, 231-342.

[62] Sun C.; Lee J.S.H.; Zhang M. Magnetic nanoparticles in MR imaging and drug delivery. *Advanced drug delivery reviews* 2008, 60, 1252-1265.

[63] Lee J.H.; Kim J.; Cheon J. Magnetic nanoparticles for multi-imaging and drug delivery. *Molecules and cells* 2013, 35, 274-284.

[64] Li X.; Li H.; Liu G.; Deng Z.; Wu S.; Li P.; Xu Z.; Xu H.; Chu P.K. Magnetite-loaded fluorine-containing polymeric micelles for magnetic resonance imaging and drug delivery. *Biomaterials* 2012, 33, 3013-3024.

[65] Qu H., Caruntu D.; Liu H., O'Connor C.J. Water-dispersible iron oxide magnetic nanoparticles with versatile surface functionalities. *American chemical society* 2011, 27, 2271-2278.

[66] Allard-Vannier E.; Cohen-Jonathan S.; Gautier J.; Herve-Aubert K.; Munnier E.; Souce M.; Legras P.; Passirani C.; Chourpa I. Pegylated magnetic nanocarriers for doxorubicin delivery: A quantitative determination of stealthiness *in vitro* and *in vivo*. *European journal of pharmaceutics and biopharmaceutics* 2012, 81, 498-505.

[67] Shen J.M.; Tang W.J.; Zhang X.L.; Chen T.; Zhang X.H. A novel carboxymethyl chitosan-based folate/Fe_3O_4/CdTe nanoparticle for targeted drug delivery and cell imaging. *Carbohydrate polymers* 2012, 88, 239-249.

[68] Cole A.J.; David A.E.; Wang J.; Galban C.J.; Yang V.C. Magnetic brain tumor targeting and biodistibution of long-circulating PEG-modified, cross-linked starch-coated iron oxide nanoparticles. *Biomaterials* 2011, 32, 6291-6301.

[69] Mu L.; Seow P.H.; Ang S.N.; Feng S.S. Study on surfactant coating of polymeric nanoparticles for controlled delivery of anticancer drug. *Colloid & polymer science* 2004, 283, 58-65.

[70] Rahimi M.; Wadajkar A.; Subramanian K.; Yousef M.; Cui W.; Hsieh J.T.; Nguyen K. In vitro evaluation of novel polymer-coated magnetic nanoparticles for controlled drug delivery. *Nanomedicine* 2010, 6 (5), 672-680.

[71] Dann, Dorfman, Herrell&Skillman. Magnetic nanoparticle composition and using the same, Labhasetwar V.; Jain T.; Leslie-Pelecky D. US2007/0264199A1. US patent 2007-11-15.

[72] Arruebo M.; Fernández-Pacheco R.; Ibarra M.R.; Santamaria J. Magnetic nanoparticles for drug delivery. *Nanotoday* 2007, 2 (3), 22-32.

[73] De Jong W.H.; Borm P.J.A. Drug delivery and nanoparticles: Applications and hazards. *International journal of nanomedicine* 2008, 3 (2), 133-149.

[74] Karlsson H.L.; Cronholm P.; Gustafsson J.; Moller L. Copper oxide nanoparticles are highly toxic: a comparison between metal oxide nanoparticles and carbon nanotubes. *Chemical research in toxicology* 2008, 21, 1726-1732.

[75] Häfeli U.O.; Riffle J.S.; Harris-Shekhawat L.; Carmichael-Baranauskas A.; Mark F.; Dailey J.P.; Bardenstein D. Cell uptake and in vitro toxicity of magnetic nanoparticles suitable for drug delivery. *Molecular pharmaceutics* 2009, 6 (5), 1417-1428.

[76] Mahmoudi M.; Simchi A.; Imani M.; Shokrgozar M.A.; Milani A.S.; Häfeli U.O.; Stroeve P. A new approach for the *in vitro* identification of the cytotoxicity of superparamagnetic iron oxide nanoparticles. *Colloids and surfaces B: Biointerfaces* 2010, 75, 300-309.

[77] Kunetsov O.A.; Brusentsov N.A.; Kuznetsov A.A.; Yurchenko N.Y.; Osipov N.E.; Bayburtskiy F.S. Correlation of the coagulation rates and toxicity of biocompatible ferromagnetic microparticles. *Journal of magnetism and magnetic materials* 1999, 194, 83-89.

[78] Park J.H.; Im K.H.; Lee S.H.; Kim D.H.; Lee D.Y.; Lee Y.K.; Kim Y.K.; Kim K.N. Preparation and characterization of magnetic chitosan particles for hyperthermia application. *Journal of magnetism and magnetic materials* 2005, 293, 328-333.

[79] Ding G.; Guo Y.; Lv Y.; Liu X.; Xu L.; Zhang X. A double-targeted magnetic nanocarrier with potential application in hydrophobic drug delivery. *Colloids and surfaces B: Biointerfaces* 2012, 91, 68-76.

[80] Carneiro M.LB.; Nunes E.S.; Peixoto R.CA.; Oliveira R.GS.; Lourenco L.HM.; Silva I.CR.; Simioni A.R.; Tedesco A.C.; Souza A.R.; Lacava Z.GM.; Bao S.N. Free rhodium (II) citrate and rhodium (II) citrate magnetic carriers as potential strategies for breast cancer therapy. *Journal of nanobiotechnology* 2011, 9:11.

[81] Kumar Dutta R.; Sahu S. Development of oxaliplatin encapsulated in magnetic nanocarriers of pectin as a potential targeted drug delivery for cancer therapy. *Results in pharma sciences* 2012, 2, 38-45.

[82] Carneiro M.L.; Peixoto R.CA.; Joanitti G.A.; Oliveira R.GS.; Telles L.AM.; Miranda-Vilela A.L.; Bocca A.L.; Vianna L.MS.; Silva I.CR.; Souza A.R.; Lacava Z.GM.; Bao S.N. Antitumor effect and toxicity of free rhodium (II) citrate and rhodium (II) citrate-loaded maghemite nanoparticles in mice bearing breast cancer. *Journal of nanobiotechnology* 2013, 11:4.

[83] Akbarzadeh A.; Samiei M.; Davaran S. Magnetic nanoparticles: preparation, physical properties, and applications in biomedicine. *Nanoscale research letters* 2012, 7, 1-13.

[84] Yoon T.J.; Yu K.N.; Kim E.; Kim J.S.; Kim B.G. Specific targeting, cell sorting, and bioimaging with smart magnetic silica core-shell nanomaterials. *Small* 2006, 2, 209-215.

[85] Yoon T.J.; Kim J.S.; Kim B.G.; Yu K.N.; Cho M.H. Multifunctional nanoparticles possessing a ″magnetic motor effect″ for drug or gene delivery. *Angewandte Chemie International Edition* 2005, 44, 1068-1071.

[86] Jeong U.; Teng X.W.; Wang Y.; Yang H.; Xia Y.N. Superparamagnetic colloids: Controlled synthesis and niche applications. *Advanced materials* 2007, 19 (1), 33-60.

[87] Hilger I.; Frühauf K.; Andrä W.; Hiergeist R.; Hergt R.; Keiser W. Heating potential of iron oxides for therapeutic purposes in interventional radiology. *Academic radiology* 2002, 9 (2), 198-202.

[88] Rosensweig R.E. Heating magnetic fluid with alternating magnetic field. *Journal of magnetism and magnetic materials* 2002, 252, 370-374.

[89] Goodwin S.; Peterson C.; Hoh C.; Bittner C. Targeting and retention of magnetic targeted carriers (MTCs) enhancing intra-arterial chemotherapy. *Journal of magnetism and magnetic materials* 1999, 194, 132-139.

[90] Casula M.F.; Jun Y.W.; Zaziski D.J.; Chan E.M. Corrias A. The concept of delayed nucleation in nanocrystal growth demonstrated for the case of iron oxide nanodisks. *Journal of the American chemical society* 2006, 128 (5), 1675-1682.

[91] Bozicevic, field & francis llp. Functionalized magnetic nanoparticles and methods of use thereof, Akhtari M.; Engel J. US 2008/0206146 A1. US patent 2008-8-28.

[92] Giri S.; Trewyn B.G.; Stellmaker M.P.; Lin V.S-Y. Stimuli-responsive controlled-release delivery system based on mesoporous silica nanorods capped with magnetic nanoparticles. *Angewandte chemie international edition* 2005, 44, 5038-5044.

[93] Lai C.Y.; Trewyn B.G.; Jeftinija D.M.; Jeftinija K.; Xu S.; Jeftinija S.; Lin V.S. A mesoporous silica nanosphere-based carrier system with chemically removable CdS nanoparticle caps for stimuli-responsive controlled release of neurotransmitters and drug molecules. *Journal of the American chemical society* 2003, 125, 4451-4459.

[94] Prijic S.; Sersa G. Magnetic nanoparticles as targeted delivery systems in oncology. *Radiol oncol* 2011, 45 (1), 1-16.

[95] Kievit F.M.; Veiseh O.; Fang C.; Bhattarai N.; Lee D.; Ellenbogen R.G. Chlorotoxin labeled magnetic nanovectors for targeted gene delivery to glioma. *ACS Nano*, 2010, 4, 4587-4594.

[96] Choi H.S.; Liu W.; Misra P.; Tanaka E.; Zimmer J.P.; Itty Ipe B.; Bawendi M.G.; Fragioni J.V. Renal clearance of quantum dots. *Nature biotechnology* 2007, 25, 1165-1170.

[97] Chen L.T.; Weiss L. The role of the sinus wall in the passage of erythrocytes through the spleen. *Blood* 1973, 41, 529-537.

[98] Fang C.; Kievit F.M.; Veiseh O.; Stephen Z.R.; Wang T.; Lee D.; Ellenbogen R.G.; Zhang M. Fabrication of magnetic nanoparticles with controllable drug loading and release through a simple assembly approach. *Journal of controlled release* 2012, 162, 233-241.

INDEX

A

B

C

D

E

F

G

H

J

K

L

M

N

O

P

Q

R

S

T